HIGH-ENERGY PHYSICS

Studies in the Natural Sciences

A Series from the Center for Theoretical Studies
University of Miami, Coral Gables, Florida

Orbis Scientiae: Behram Kursunoglu, *Chairman*

Recent Volumes in this Series

Volume 8 **PROGRESS IN LASERS AND LASER FUSION**
Edited by Arnold Perlmutter, Susan M. Widmayer, Uri Bernstein, Joseph Hubbard, Christian Le Monnier de Gouville, Laurence Mittag, Donald Pettengill, George Soukup, and M. Y. Wang

Volume 9 **THEORIES AND EXPERIMENTS IN HIGH-ENERGY PHYSICS**
Edited by Arnold Perlmutter, Susan M. Widmayer, Uri Bernstein, Joseph Hubbard, Christian Le Monnier de Gouville, Laurence Mittag, Donald Pettengill, George Soukup, and M. Y. Wang

Volume 10 **NEW PATHWAYS IN HIGH-ENERGY PHYSICS I**
Magnetic Charge and Other Fundamental Approaches
Edited by Arnold Perlmutter

Volume 11 **NEW PATHWAYS IN HIGH-ENERGY PHYSICS II**
New Particles—Theories and Experiments
Edited by Arnold Perlmutter

Volume 12 **DEEPER PATHWAYS IN HIGH-ENERGY PHYSICS**
Edited by Arnold Perlmutter, Linda F. Scott, Mou-Shan Chen, Joseph Hubbard, Michel Mille, and Mario Rasetti

Volume 13 **THE SIGNIFICANCE OF NONLINEARITY IN THE NATURAL SCIENCES**
Edited by Arnold Perlmutter, Linda F. Scott, Mou-Shan Chen, Joseph Hubbard, Michel Mille, and Mario Rasetti

Volume 14 **NEW FRONTIERS IN HIGH-ENERGY PHYSICS**
Edited by Arnold Perlmutter, Linda F. Scott, Osman Kadiroglu, Jerzy Nowakowski, and Frank Krausz

Volume 15 **ON THE PATH OF ALBERT EINSTEIN**
Edited by Arnold Perlmutter and Linda F. Scott

Volume 16 **HIGH-ENERGY PHYSICS IN THE EINSTEIN CENTENNIAL YEAR**
Edited by Arnold Perlmutter, Frank Krausz, and Linda F. Scott

Volume 17 **RECENT DEVELOPMENTS IN HIGH-ENERGY PHYSICS**
Edited by Arnold Perlmutter and Linda F. Scott

Volume 18 **GAUGE THEORIES, MASSIVE NEUTRINOS, AND PROTON DECAY**
Edited by Arnold Perlmutter

Volume 19 **FIELD THEORY IN ELEMENTARY PARTICLES**
Edited by Arnold Perlmutter

Volume 20 **HIGH-ENERGY PHYSICS**
Edited by Stephan L. Mintz and Arnold Perlmutter

Volume 21 **INFORMATION PROCESSING IN BIOLOGICAL SYSTEMS**
Edited by Stephan L. Mintz and Arnold Perlmutter

---

fm

(HIGH-ENERGY PHYSICS)

In Honor of P.A.M. Dirac
in his Eightieth Year

Chairman

Behram Kursunoglu

Editors

Stephan L. Mintz

Florida International University
Miami, Florida

and

Arnold Perlmutter

Center for Theoretical Studies
University of Miami
Coral Gables, Florida

PLENUM PRESS • NEW YORK AND LONDON

Library of Congress Cataloging in Publication Data

Orbis Scientiae (1983: Miami, Fla.)
High-energy physics.

(Studies in the natural sciences; v. 20)
"Proceedings of the first half of the 20th annual Orbis Scientiae, dedicated to P. A. M. Dirac's 80th year, held January 17–21, 1983, in Miami, Florida"—T.p. verso.
Includes bibliographical references and index.
1. Particles (Nuclear physics)—Congresses. 2. Astrophysics—Congresses. 3. Dirac, P. A. M. (Paul Adrien Maurice), 1902– . I. Kursunoglu, Behram, 1922– II. Mintz, Stephan L. III. Perlmutter, Arnold, 1928– . IV. Dirac, P. A. M. (Paul Adrien Maurice), 1902– . V. Title. VI. Series.
QC793.O7 1983 539.7'54 85-12180
ISBN 0-306-42070-8

Proceedings of the first half of the 20th Annual Orbis Scientiae,
dedicated to P. A. M. Dirac's 80th year,
held January 17–21, 1983, in Miami, Florida

A Division of Plenum Publishing Corporation
233 Spring Street, New York, N.Y. 10013

Printed in the United States of America

The Editors Dedicate this Volume
to the Memory of
Paul Adrien Maurice Dirac
(1902–1984)

PREFACE

This volume contains the greater part of the papers submitted to the High Energy Physics portion of the 1983 Orbis Scientiae, then dedicated to the eightieth year of Professor P. A. M. Dirac. Before the volume could be published, Professor Dirac passed away on October 20, 1984, thereby changing the dedication of this volume, and its companion, on Information Processing in Biology, to his everlasting memory.

Since 1969, Professor Dirac had given the opening address at each of these conferences. He was unable to prepare a manuscript of his last paper in 1983. His impact on science already has been enormous. The consequences of his thought and work for future developments are incalculable.

Regrettably, Professor Dirac's last appearance at this series of conferences, begun in 1964 as the Coral Gables Conference on Symmetry Principles at High Energy, coincided with the twentieth, and the last of these. The work and expense involved in organizing them and preparing the proceedings have come to far exceed the physical capabilities and the support received by the Center for Theoretical Studies for this program. The delayed appearance of these proceedings, for which the editors humbly apologize, is a manifestation of the inadequate support. On the other hand, the organizers and editors thank the many distinguished participants who, over the years, made these meetings exciting and productive arenas for the dissemination of ideas in high energy physics and related fields.

The last Orbis Scientiae was (as it was often in the past) shared by two frontier fields: high energy physics and information processing in biology, demonstrating the universality of scientific principles and goals. The interaction amongst scientists of diverse interests can only enhance the fruitfulness of their efforts. The editors take pride in the modest contribution of Orbis Scientiae towards this goal.

It is quite possible that, in the near future, Orbis Scientiae may be revived with a different format, perhaps in the form of smaller, intensive seminars on more specific subjects.

It is a pleasure to acknowledge the typing of these proceedings by Rogelio Rodriguez and Helga Billings, and the customary excellent supervision by the latter. The efficient preparation and organization of the conference was due largely to the skill and dedication of Linda Scott. As in the past, Orbis Scientiae 1983 received nominal support from the United States Department of Energy and the National Science Foundation.

The Editors
Coral Gables, Florida
April, 1985

CONTENTS

COSMOLOGICAL AND ASTROPHYSICAL IMPLICATIONS OF MAGNETIC MONOPOLES

Edward W. Kolb†

Theoretical Division

Los Alamos National Labratory

Los Alamos, NM 87545

ABSTRACT

Among Dirac's many contributions to modern physics is the idea that charge quantization is natural in a theory with magnetic monopoles. The existence of magnetic monopoles would have drastic effects on the evolution of the universe, on galactic magnetic fields, and perhaps on the x-ray luminosity of neutron stars. In this talk I will review some astrophysical implications of massive monopoles.

In 1931, Dirac showed that the quantization of electric charge follows in any theory which also includes magnetic charge.[1] In such theories, the electric charge, e, and magnetic charge, g, are quantized according to the "Dirac quantization condition"

$$eg = n/2 \qquad (n \text{ integer}). \tag{1}$$

As originally introduced by Dirac, the monopole was simply a new dynamical degree of freedom to be added to the system,

†Work supported in part by the Department of Energy.

and all the monopole properties, such as mass, spin, electric charge, etc., were free parameters.

This situation changed in 1974, when t' Hooft and Polyakov [2, 3] demonstrated that monopoles <u>must</u> occur in certain classes of gauge theories. Among the classes of gauge theories with monopoles are the grand unified theories (GUTS), which attempt to unify the strong and electroweak interactions. In these theories, all the monopole properties are calculable. In this talk, I will consider only the monopoles in grand unified models based on SU(5). The specialization to SU(5) is for convenience only; the same conclusions would obtain in any GUT.

The mass of the GUT monopole is related to the scale of the breaking $G \rightarrow G' \otimes U(1)$, where G is the grand group. In SU(5), this would correspond to the mass scale for the breaking $SU(5) \rightarrow SU(3) \otimes SU(2) \otimes U(1)$. The limit on the proton lifetime requires this scale to be greater than about 10^{14} GeV. The mass of the monopole is roughly this scale divided by α, or about 10^{16} GeV. This enormous mass 10^{-8} g) prevents creation of monopoles in accelerators. Only in the early universe can the energy be found to create the monopoles.

There are two sources of monopole creation in the early universe; thermal production, and spontaneous production in the phase transition.[4,5] As spontaneous production dominates in the standard cosmology I will ignore thermal production.

The key ingredient in spontaneous production is the fact that at high temperatures, thermal effects of the ambient background gas should have restored the symmetry. That is, at temperatures much greater than 10^{15} GeV, $\langle\phi\rangle = 0$, where $\langle\phi\rangle$ is the vacuum expectation value of Higgs field responsible for the breaking $SU(5) \rightarrow SU(3) \otimes SU(2) \otimes U(1)$. Since monopoles exist as knots in $\langle\phi\rangle$, in the symmetric phase ($\langle\phi\rangle = 0$) there are no monopoles. At some critical temperature, $T_c \cong 10^{15}$ GeV, spontaneous symmetry breaking occurs, and monopoles will be formed. The number of monopoles formed depends on the distance over which the Higgs field can be correlated, which in turn, depends upon the size of the causal horizon.

The causal horizon, d_H, in the standard big bang model is given by

$$d_H = 2t \cong m_{P\ell}/10T^2 \quad , \tag{2}$$

where t is the time since the bang, which can be expressed in terms of temperature (T) and the Planck mass ($m_{P\ell}$) in the standard model. The Higgs field cannot be correlated over the distances larger than d_H, and the finiteness of the correlation length leads to the production of monopoles. The number of monopoles produced may be calculated by considering three points, A, B, and C separated by the distance greater than d_H. At each point, we can represent the orientation of $\langle\phi\rangle$ in group space by a vector. As the distances $\overline{AB}$, $\overline{AC}$, $\overline{BC}$ are greater than the causal horizon, the direction in group space of $\langle\phi\rangle$ at the three points are independent. There exists a finite probability, about 10%, that any given orientation of the vectors will correspond to a nontrivial topology, i.e., correspond to a monopole.[6] Therefore, the expectation is that on average one monopole per 10 horizon volumes should be created (and an equal number of antimonopoles). This leads to the creation of a number density of monopole-antimonopoles pairs of

$$n_M = n_{\bar{M}} \cong 0.1\, d_H^{-3} \cong 10^3\, T_c^6/m_{P\ell}^3 \quad . \tag{3}$$

It is convenient to compare this number density to the photon number density at $T = T_c$, $n_\gamma \cong T_c^3$:

$$\frac{n_M}{n_\gamma} = \frac{n_{\bar{M}}}{n_\gamma} \cong 10^3 \frac{T_c^3}{m_{P\ell}^3} \cong 10^{-9} \qquad (T_c = 10^{15}\ \text{GeV}) . \tag{4}$$

After creation of the monopoles, monopole-antimonopole pairs can annihilate. This annihilation causes the monopole density to decrease as (ignoring monopole-antimonopole creation in collisions)

$$\frac{dn_M}{dt} = \frac{dn_{\bar{M}}}{dt} = -\langle n_M n_{\bar{M}}\, \sigma_A |v|\rangle - 3\,\frac{\dot{R}}{R}\, n_M \quad , \tag{5}$$

where σ_A is the $M\bar{M}$ annihilation cross section. The classical calculation of σ_A involves monopole-antimonopole capture into a bound state, with subsequent emission of radiation until the eventual annihilation.[4, 5, 7] The solution to (5) results in a "freeze out" density of monopoles of about

$$\frac{n_{\bar{M}}}{n_\gamma} = \frac{n_M}{n_\gamma} \cong 10^{-10} \quad . \tag{6}$$

The ratio in (6) is roughly constant during the subsequent evolution of the universe. Therefore, today the monopoles would contribute an energy density of

$$\rho_{M\bar{M}} = m_M n_M \cong 10^{-10} \; m_M n_\gamma \quad . \tag{7}$$

There are about 400 photons cm^{-3} in the universe today, and if the monopole mass is 10^{16} GeV (10^{-8} g), the monopoles would contribute an energy density of

$$\rho_{M\bar{M}} \cong 10^{-15} \text{ g cm}^{-3} \cong 10^{14} \rho_c \quad , \tag{8}$$

where ρ_c is the critical density. Since observationally we know the total energy density of the universe must be less than $2\rho_c$, the energy density in (8) is about 14 orders of magnitude too large. Therefore, there must be some way either to avoid making so many monopoles, or to efficiently annihilate the monopoles.

The obvious lesson is that we are missing something very fundamental in the simple picture of the early universe we have just discussed. Below we discuss some proposals that have been made to alleviate the monopole problem.

The monopoles are formed in the phase transition; if one could avoid the phase transition, monopoles would not be produced. One possible way to avoid the phase transition is to assume that the universe had a maximum temperature, T_{MAX}, less than the critical temperature, T_c. This assumption has the unwanted feature that baryon production is difficult, since temperatures of the order

$T_c \cong m_X$ are necessary to make baryons. There may be a narrow range for T_{MAX}, where $T_{MAX} \lesssim T_c$, and T_{MAX} is still large enough to produce baryons; but unless there is some physical reason for T_{MAX} to be in that narrow range, this solution cannot be taken seriously. One possible physical reason for $T_{MAX} \lesssim T_c$ can arise in grand unified models which are not asymptotically free.[8] Such asymptotically non-free models are considered ugly on aesthetic grounds, but quite naturally arise in complicated GUTS.[8]

Another possible way to prevent the universe from being in the symmetric state, is to assume that there is a large fermion asymmetry that prevents high temperature restoration of symmetry.[9] Such an initial large fermion number may have further implications for cosmology, e.g., for the neutrino number of the universe.[10]

One possibility in enhancing annihilation is to assume that the monopoles are clumped, and n_M in eq. (5) is larger than the value obtained in uniform expansion. The gravitational attraction of monopoles provides just such a mechanism. The possibility that monopoles form bound objects and annihilate was suggest by Preskill,[5] and examined in detail by several groups.[7, 11] All have concluded that the mechanism of gravitational clumping cannot solve the monopole problem.

Another possibility is that when the monopoles and antimonopoles form, they are connected by strings, and are confined.[12] A basic assumption in eq. 5 is that the initial states are uncorrelated in the scattering. If monopoles are connected to antimonopoles by strings, the initial state is correlated, and the annihilation rate will be much enhanced. Although the physics behind the motivation is somewhat uncertain, there are at least some reasons to believe in monopole confinement since monopole confinement could be an artifact of the non-Abelian nature of the monopole.

The number of monopoles created in the transition is proportional to $T_c^3/m_{P\ell}^3$. If the critical temperature is less than about 10^{10} GeV, an acceptable monopole density would result. One method of postponing the phase transition involves supercooling,[13, 14] and was the original motivation of the inflationary universe. Another possibility is an intermediate symmetry as proposed by Langacker and Pi,[15] where the U(1) appearance is postponed until $T \lesssim 10^{10}$ GeV.

In conclusion, the dearth of monopoles in the universe gives us very important information about the early universe. Unfortunately, it is not clear how to interpret the information. Perhaps one of the scenarios for monopole suppression discussed above is the solution to the "monopole problem." but it is just as likely that the solution has not yet been found. Some of the solutions discussed above get rid of all the monopoles. Some of the solutions get rid of some of the monopoles. The obvious next step is to look for monopoles today. The next section discusses limits on the present monopole flux, both astrophysical limts and experimental limits.

So far we have seen that massive magnetic monopoles are predicted to exist in grand unified theories, and should have been copiously produced in the big bang. In fact, the predicted density of monopoles is so large that some mechanism must either suppress their production, or efficiently reduce their number after creation. If monopoles could be detected, knowledge of the present monopole flux would help in determining the suppression mechanism.

Before discussing terrestrial experiments, it is useful to discuss some astrophysical limits. The first limit comes from the mass density of the universe. The "critical density" for closure of the universe is (H_0 is the Hubble constant) $\rho_c \cong 2 \times 10^{-29}$ g cm^{-3}. We know observationally that the energy density of the universe is less than $2\rho_c$. This implies that the average member density of monopoles, $\langle n_M \rangle$, is limited by requiring that the energy density of the monopoles, $\rho_M = \langle n_M \rangle$ is less than $2\rho_c$

$$\langle n_M \rangle \lesssim 4 \times 10^{-21} \frac{10^{16}\text{GeV}}{m_M} \text{ cm}^{-3} \quad . \tag{9}$$

In order to compare with experiments, we must form a flux from the density. If we assume the monopoles are completely uniform throughout the universe, they would be cold, and the relevant velocity would be the velocity of our galaxy relative to the cold monopoles. This velocity is our peculiar velocity relative to the 3K background radiation,[16] $v_M \cong 10^{-3}$ c. Therefore, the flux of monopoles in the universe must be

$$\begin{aligned} \langle F_M \rangle_U &\lesssim \langle n_M \rangle_U \, v_M \\ &\lesssim 10^{-14} \frac{10^{16}\ \text{GeV}}{m_M} \frac{v_M}{10^{-3}} \ \text{cm}^{-2}\ \text{s}^{-1}\ \text{sr}^{-1} \ . \end{aligned} \tag{10}$$

It is perhaps more reasonable to assume that the monopoles are clumped and that their density in our galaxy is enhanced over the average density throughout the universe, as are the protons. In that case, the relevant mass density limit is roughly $10^6\ \rho_c$, and the relevant velocity would be the galactic virial velocity, again about 10^{-3} c. The limit in this case is

$$\langle F_M \rangle_G \lesssim 10^{-9} \frac{10^{16}\text{GeV}}{m_M} \ \text{cm}^{-2}\ \text{s}^{-1}\ \text{sr}^{-1} \ . \tag{11}$$

The best astrophysical limit comes from the observation of large-scale galactic magnetic fields. The original argument is due to Parker.[17] He argued that if the galactic magnetic fields are due to currents, then monopoles drain energy from the field as they are accelerated. The rate of extracting energy from the magnetic field is $F_M B_G$ where B_G is the galactic magnetic field ($B_G \cong 10^{-6}$ gauss). The energy in the magnetic field is $B_G^2/8\pi$, and this energy will be drained on a timescale $\tau_B = B^2/32\pi^2\ F_M B$. In order for magnetic fields to survive, this timescale must be less than the timescale for regeneration of the magnetic field, which is the order of the galactic rotation time, about 10^8 years.[17] The requirement $\tau_B \gtrsim 10^8$ years requires

$$F_M \lesssim 10^{-16}\ \text{cm}^{-2}\ \text{s}^{-1}\ \text{sr}^{-1} \ . \tag{12}$$

Equation (12) has become known as the "Parker limit." More detailed studies (18) have closed several loopholes in the simple argument given above, and have concluded that the limit (12) is probably good to an order of magnitude for reasonable values of the monopole mass. It is possible to evade the Parker limit if the monopoles themselves are the sources of the magnetic field.[18, 19] This scenario is hard to imagine, since it needs large magnetic charge fluctuations. However, it potentially weakens the Parker limit by about 5 orders of magnitude.

The uncertainty in the astrophysical limits is easily matched by the uncertainties in terrestrial searches. The largest source of uncertainty comes from the difficulty in calculating the energy loss from slowly moving monopoles. For a discussion of the difficulties of such a calculation, see Ahlen and Kinoshita.[20] They find an energy loss for massive, slowly moving monopoles in silicon of $dE/dx \cong 50\ \beta$ GeV cm^{-1} for $\beta \leq 10^{-2}$, where β is the monopole velocity.

All reported experiments that detect monopoles by ionization depend on calculations of the monopole ionization. Although different values for the ioniztion, it is still interesting to compare some quoted limits:

$$F_M \lesssim 10^{-10}\ \mathrm{cm}^{-2}\ \mathrm{s}^{-1}\ \mathrm{sr}^{-1} \qquad (\beta \gtrsim 3 \times 10^{-3}) \qquad (13a)$$

$$F_M \lesssim 3 \times 10^{-11}\ \mathrm{cm}^{-2}\ \mathrm{s}^{-1}\ \mathrm{sr}^{-1} \qquad (3 \times 10^{-4} \leq \beta \leq 10^{-3}) \qquad (13b)$$

$$F_M \lesssim 10^{-11}\ \mathrm{cm}^{-2}\ \mathrm{s}^{-1}\ \mathrm{sr}^{-1} \qquad (2 \times 10^{-4} \leq \beta \leq 3 \times 10^{-2}) \qquad (13c)$$

$$F_M \lesssim 2 \times 10^{-12}\ \mathrm{cm}^{-2}\ \mathrm{s}^{-1}\ \mathrm{sr}^{-1} \qquad (10^{-2} \lesssim \beta) \qquad (13d)$$

$$F_M \lesssim 3 \times 10^{-13}\ \mathrm{cm}^{-2}\ \mathrm{s}^{-1}\ \mathrm{sr}^{-1} \qquad (10^{-3} \leq \beta \leq 10^{-2})\ , \qquad (13e)$$

where the limits quoted in (13a) ··· (13e) are given in refs. 21 ··· 25, respectively.

The cleanest way to look for monopoles is the superconducting search done by Cabrera.[26] This method is based on the change in the microscopic quantum state of a superconducting ring when a magnetic charge passes through the ring. The magnetic flux through the superconducting ring changes the current, which is detected by a SQUID. This method is clean, since it is independent of the monopole mass, monopole velocity, monopole electric charge, and energy loss calculations. Cabrera has reported a limit [26]

$$F_M \lesssim 6 \times 10^{-10}\ \mathrm{cm}^{-2}\ \mathrm{s}^{-1}\ \mathrm{sr}^{-1} \quad , \tag{14}$$

and a candidate event. If we assume the candidate event was the actual detection of a magnetic monopole, the flux of 6×10^{-10} cm^{-2} s^{-1} sr^{-1} represents the actual monopole flux.

Although one event does not give unequivocal evidence for the existence of magnetic monopoles, it is interesting to assume this event is the result of a flux given in (14) and examine the consequences. First, it is obviously in conflict with the ionization searches discussed above. This would mean that either the monopoles are very slow (β << galactic virial velocity) or the ionization calculations are in serious error. The Cabrera flux is also in serious disagreement with the Parker limit. Either we do not understand galactic magnetic fields, or the monopole flux on earth is much different than the average galactic flux. The latter possibility has been examined by Dimopoulos, Glashow, Purcell and Wilczek.[27] They suggested that there is a "cloud" of monopoles orbiting the sun, which enhances the monopole flux at the earth by a factor of 10^5 relative to the average galactic flux. In this manner it may be possible to reconcile the Cabrera flux with the Parker limit.

In conclusion, there is one hint that monopoles exist in our galaxy. If this hint proves correct, it may have drastic implications for our understanding of galactic magnetic fields.

Two dramatic predictions of GUTS are proton decay, and the existence of magnetic monopoles. Recent model calculations by Rubakov[28], Callan[29], and Wilczek[30] have suggested that proton decay in the presence of a monopole occurs at a much faster rate than one might guess on dimensional grounds.

A naive dimensional estimate of the cross section of $pM \to MX$, where X denotes the proton decay products, might be $\sigma_{\Delta B} = m_M^{-2}$, where m_M is the monopole mass. However, Rubakov and Callan have argued that the cross section is <u>independent</u> of the mass of the monopole, and it is related instead to the mass scale of color confinement, about 1 GeV. We shall parameterize the cross section for monopole-induced proton decay by a mass scale Λ:

$$\sigma_{\Delta B} = \frac{\pi}{\Lambda^2} \cong \frac{10^{-27}}{\Lambda^2_{GeV}} \; cm^2 \quad , \tag{15}$$

where $\Lambda_{GeV} = \Lambda/1$ GeV. If the conjecture of Callan and Rubakov is correct, then $\Lambda_{GeV} \cong 1$.

There are two obvious questions to be asked about monopole-induced proton decay. First, does it open up new methods for detecting cosmic monopoles. Second, does monopole-induced proton decay have any important astrophysical consequences. The answer to both questions is yes!

If monopoles induce proton decay while traversing a proton decay detector, then proton decay detectors would make the most sensitive monopole telescopes. A preliminary study by the Irvine-Michigan-Brookhaven proton decay collaboration has concluded that with minor modifications, the detector would have a sensitivity to monopole flux of about 10^{-15} cm^{-2} s^{-1} sr^{-1}, if $\sigma_{\Delta B} \cong 10^{-27}$ cm^2.[31] This is several orders of magnitude better than the limits in eqs. (13).

If monopoles induce proton decay, they would have a drastic effect on the x-ray luminosity of neutron stars.[32, 33] By using the limit on the neutron star luminosity, it is possible to place a limit on the product of the galactic monopole flux and baryon violating cross section, which suggests that the phenomena of monopole-induced proton decay is not likely to be detected terrestrially.

Neutron stars are believed to be plentiful in our galaxy,[34] and to be as old as the galaxy itself. During the lifetime of the neutron star, the neutron star would have been subjected to a galactic monopole flux F_M. This resulted in a number of neutron star/monopole collisions given by

$$N_M = \frac{2\pi}{3} F_M A_c t_{NS} , \tag{16}$$

where t_{NS} is the lifetime of the neutron star, and A_c is the "capture" area given by

$$A_c = R_{NS}^2 \frac{1 + 2M_{NS} G/v_M^2 R_{NS}}{1 - R_S/R_{NS}} . \tag{17}$$

In eq. (17), M_{NS} and R_{NS} are the mass and radius of the neutron star, respectively, v_M is the galactic velocity of the monopole, and R_S is the Schwarzschild radius of the neutron star. Putting in values of $M_{NS} = 1\ M_O$, $R_{NS} = 10^6$ cm,

$$N_M = 10^{36} F_M(\text{cm}^{-2}\ \text{s}^{-1}\ \text{sr}^{-1}) . \tag{18}$$

The monopole will be trapped in the neutron star, mainly through energy loss in scattering with electrons.[20, 23] This results in an energy loss of

$$\frac{dE}{dx} \cong \beta 10^{11}\ \text{GeV cm}^{-1} , \tag{19}$$

where β is the monopole velocity in the star ($\beta \cong 10^{-1}$ is escape velocity). If $\beta \gtrsim 10^{-1}$, the monopole will lose about 10^{16} GeV traversing the star (if $\beta \lesssim 10^{-1}$ the monopole must be trapped). The initial energy of the monopole in the galaxy was

$$\begin{aligned} E_O &= \frac{1}{2} m_M v_M^2 \\ &\cong 5 \times 10^9 \frac{m_M}{10^{16}\text{GeV}} \left[\frac{v_M}{10^{-3}}\right]^2 \text{GeV} . \end{aligned} \tag{20}$$

Therefore, if the monopoles have typical virial velocities of 10^{-3}, monopoles of mass less than 10^{22} GeV (10^3 $m_{P\ell}$!) will be trapped.

The monopoles trapped in the star catalyze nucleon decay, and release energy at a rate

$$L_M = m_N \, n_N \, \sigma_{\Delta B} \, |v| \qquad (21)$$

$$\cong 8.5 \times 10^{17} \, \Lambda_{GeV}^{-2} \text{ erg s}^{-1} \text{ monopole}^{-1}.$$

where m_N is the nucleon mass, n_N is the nucleon density, and we have used $|v| \cong 10^{-1}_{c}$, the nucleon Fermi velocity.

Combining eqs. (18) and (21), the energy produced in the neutron star by monopole-induced nucleon decay is

$$L = 2 \times 10^{54} \, \Lambda_{GeV}^{-2} \, F_M(\text{cm}^{-2} \text{ s}^{-1} \text{ sr}^{-1}) \text{ erg s}^{-1} \quad . \qquad (22)$$

This luminosity is the total luminosity, which in general may be emitted as photons and neutrinos. The best limit on the photon luminosity comes from recent surveys for serendipitous x-ray sources.[35] These surveys could detect an x-ray luminosity of $L_\gamma = 10^{31}$ erg s^{-1} at a distance of 1 kpc.[35] Estimates based on the current pulsar birth rate in the solar neighborhood predict there should be an old neutron star density of $n_{NS} \gtrsim 4 \times 10^{-3}$ pc^{-3}. Therefore, it is reasonable as an extremely conservative estimate, to take 10^{31} erg s^{-1} as the maximum luminosity of old neutron stars.

The relationship between the x-ray luminosity and the total luminosity depends on the structure of the neutron star. The smallest L^γ/L^ν ratio results for zero surface magnetic field, and for pion condensation in the neutron star interior. In this case, $L \lesssim 10^{31}$ erg s^{-1} corresponds to a total luminosity of about 10^{33} erg s^{-1}.[36] It is also important to note that a photon luminosity of 10^{31} erg s^{-1} corresponds to a surface temperature of $T_S = 0.04$ keV. About 80% of the energy radiated at this temperature will be in the window of the Einstein satellite.

A maximum total luminosity of 10^{33} erg s^{-1} requires

$$F_M(\mathrm{cm}^{-2}\ \mathrm{s}^{-1}\ \mathrm{sr}^{-1})\ \Lambda_{\mathrm{GeV}}^{-2} < 5 \times 10^{-22} \quad . \qquad (23)$$

Therefore, if monopoles catalyze proton decay at a strong rate ($\Lambda_{\mathrm{GeV}} \doteq 1$) the flux limit of eq. (23) is about six orders of magnitude better than the Parker limit discussed in the previous section. This limit is also six order of magnitude below the possible limit for detection reported by the IMB group.

Conversely, if proton decay detectors do see monopoles, something very basic is very wrong in our understanding of neutron stars.

Not only do magnetic monopoles provide a natural explanation for charge quantization, but they must exist in grand unified theories. In such theories, the monopole is very massive, and the early universe is probably the only source for production.

The early universe should have provided a plentiful supply of monopoles. In fact, the simplest prediction one might make results in a monopole density 10^{14} orders of magnitude too large. This important cosmological probe of the early universe tells us that either the simple picture of unification, or the simple picture of the early universe is seriously wrong. There have been several proposals to solve the monopole problem; they all have important implications for cosmology, and could potentially alter our picture of the early universe. Detection of a monopole would be strong evidence that the universe was once at a temperature greater than 10^{15} GeV.

There have been several recent terrestrial searches for massive magnetic monopoles. One experiment[26] can be interpreted as detection of a monopole. If this experiment is confirmed, then the flux of magnetic monopoles would be many orders of magnitude too large for us to understand the existence of galactic magnetic fields.

There are theoretical indications that the cross section for monopole-induced proton decay is a strong cross section. If this is true, then a galactic monopole flux greater than 10^{-22} cm $^{-2}$ s^{-1} sr^{-1} would result in an enormous x-ray luminosity for neutron stars. Therefore, a detection of monopoles on earth would not only mean that galactic magnetic fields do not exist, but neutron stars do not exist.

Not only do magnetic monopoles provide a probe of particle physics, but potentially are a great source of information in astrophysics.

REFERENCES

1. P.A.M. Dirac, Proc. Roy. Soc. 133, 60 (1931).
2. G. 't Hooft, Nucl. Phys. B79, 276 (1974).
3. A.M. Polyakov, JETP Lett. 20, 194 (1974).
4. Ya. B. Zel'dovich and M. Yu. Khlopov, Phys. Lett. 79B, 239 (1978).
5. J. P. Preskill, Phys. Rev. Lett. 43, 1365 (1979).
6. T. W. B. Kibble, J. Phys. A9, 1387 (1976).
7. D. A. Dicus, D. N. Page, and V. L. Teptlitz, Phys. Rev. D26, 1306 (1982).
8. J. A. Harvey, E. W. Kolb, and S. Wolfram, Phys. Rev. D27, 315 (1983).
9. A. Linde, Phys. Rev. D14, 3345 (1976).
10. J. A. Harvey and E. W. Kolb, Phys. Rev. D24, 2090 (1981).
11. T. Goldman, E. W. Kolb, and D. Toussaint, Phys. Rev. D23, 867 (1981); J. N. Fry, Ap. J. 246, L93 (1981).
12. G. Lazarides and Q. Shafi, Phys. Lett. 94B, 149 (1980).
13. A. H. Guth and S.-H. H. Tye, Phys. Rev. Lett. 44, 631 (1980).
14. M. B. Einhorn, D. L. Stein, and D. Toussaint, Phys. Rev. D21, 3295 (1980).
15. P. Langacker and S.-Y. Pi, Phys. Rev. Lett. 45, 1 (1980).
16. G. F. Smoot and P. M. Lubin, Ap. J. 234, L83 (1979); Fabbri, et al., Phys. Rev. Lett. 44, 1563 (1980); Boughn, Cheng, and Wilkinson, Ap. J. 243, L113 (1981).
17. E. N. Parker, Ap. J. 160, 383 (1970); Ap. J. 163, 225 (1971); Ap. J. 166, 295 (1971).
18. M. S. Turner, E. N. Parker, and T. J. Bogdan, Phys. Rev. D26, 1296 (1982).

19. E. E. Salpeter, S. L. Shapiro, and I. Wasserman, Phys. Rev. Lett. 49, 1114 (1982).
20. S. P. Ahlen and K. Kinoshita, Phys. Rev. 26D, 2347 (1982); see also S. D. Drell, et al. SLAC Report, unpublished.
21. J. K. Sokolowski and L. R. Sulak, University of Michigan Report, unpublished.
22. J. D. Ullman, Phys. Rev. Lett. 47, 289 (1981).
23. D. E. Groom, et al., University of Utal Report, unpublished.
24. R. Bonarelli, et al. Phys. Lett. 112B, 102 (1982).
25. J. Bartelt, et al., Argonne Report, unpublished.
26. B. Cabrera, Phys. Rev. Lett. 48, 1378 (1982).
27. S. Dimopoulos, S. L. Glashow, E. M. Purcell, and F. Wilczek, Nature 298, 824 (1982).
28. V. A. Rubakov, Nucl. Phys. B203, 311 (1982); Zheft Pis'ma 33, 658 (1981).
29. C. G. Callan, Phys. Rev. D25, 2141 (1982); D26, 2058 (1982); Nucl. Phys. B204 (1982).
30. F. Wilczek, Phys. Rev. Lett. 48 1146 (1982).
31. S. Errede, private communication.
32. E. W. Kolb, S. A. Colgate, and J. A. Harvey, Phys. Rev. Lett. 49, 1373 (1982).
33. S. Dimopoulos, J. P. Preskill, and F. Wilczek, Phys. Lett. B.
34. D. Q. Lamb, F. K. Lamb, and D. Pines, Nature 246, 52 (1973); J. G. Hills, Ap. J. 219 (1978); 240, 242 (1980).
35. F. S. Cordova, K. O. Mason, and J. E. Nelson, Ap. J. 245, 609 (1981); G. A. Reichert, K. O. Mason, J. R. Thorstensen, and S. Bowyer, to be published.
36. K. A. Van Riper and D. Q. Lamb, Ap. J. 244, L13 (1981).

THE ART OF DOING PHYSICS IN DIRAC'S WAY

F. Rohrlich

Syracuse University

Syracuse, New York, 13210

It is a great pleasure for me to have the opportunity to contribute to this volume in recognition of Professor Paul A.M. Dirac, at the occasion of his eightieth year. While I have not had the privilege to be among his students, I have been a student of his books and papers all my life as a physicist. I first encountered his <u>Principles of Quantum Mechanics</u>[1] as a graduate student at Harvard. (Before that time I had studied engineering and had not been exposed to a course in quantum mechanics.) This remarkable book made a deep impression on me. Its elegant style and its direct and straight reasoning will forever remain an example of exposition at its best.

In the following pages I want to dwell on certain aspects of Dirac's work that are not often discussed and that I consider of great importance for a proper appreciation of his tremendous contributions to physics.

Beauty and Simplicity

The great physicist Paul Ehrenfest is said to have given the advice: "Die Eleganz soll man den Schneidern uberlassen" (elegance should be left for the tailors.) I do not believe that Dirac subscribes to this dictum. He has indicated that on numerous occasions: a correct scientific theory is both simple and beautiful. And elegance is largely a combination of these two. In his view simplicity and beauty can be used as criteria with the latter being the important one.

"It often happens that the requirements of simplicity and beauty are the same, but where they clash, the latter must take precedence."[2]

Holding simplicity and beauty to be essential attributes of scientific truth is not mere talk. He has put this belief to work in his attitude toward various branches of theoretical physics; he has chosen his research problems often after being motivated by this belief.

When he was satisfied with nonrelativistic quantum mechanics, he expressed this in his Bakerian Lectures[3] as follows: "The formalism...is so natural and beautiful as to make one feel sure of its correctness as the foundations of the theory."

But by the same criteria, he was dissatisfied with quantum electrodynamics; he wrote in 1951[4]: "Recent work by Lamb, Schwinger, Feynman and others has been very successful in setting up rules for handling the infinities and subtracting them away, so as to leave finite residues which can be compared with experiments, but the resulting theory is an ugly and incomplete one, and cannot be considered as a satisfactory solution of the problem of the electron."

Since he was one of the main architects of both nonrelativistic quantum mechanics and relativistic quantum electrodynamics, this judgment was not a criticism of other people's work but was rather more like an artist's view on one of his own creations.

Dirac has expressed similar opposition to renormalization theory repeatedly. He regards it as a set of rules for computations. But he did not leave it at that; he made several concerted efforts to improve on it. He attempted to develop quantum electrodynamics from the beginning with entirely new physical ideas at several occasions.[4-7] But he has not succeeded, so far, to his satisfaction.

Sensing the Right Answers

Ten years ago, at the occasion of Dirac's seventieth birthday, I gave a lengthy review of the history and present state of the theory of a classical point charge.[8] I will therefore mention here only one new development since that review. It provides a beautiful illustration of Dirac's keen sense for the correct answer even before experimental or theoretical evidence becomes available.

In 1938 Dirac published his well-known paper on the classical point electron in which he derived what has become known as the Lorentz-Dirac equation.[9] He gave four reasons for his assumption that the classical electron be a point rather than an extended charge.

The previous model of the electron due to Abraham and Lorentz was a charged sphere whose radius a was chosen so that its mass was entirely of electromagnetic origin,

$$m = m_{elm} \quad , \tag{1}$$

$$m_{elm}c^2 = f\,\frac{e^2}{2a} \quad . \tag{2}$$

Here f is a numerical factor of order 1 which depends on the structure of the electron, i.e. on the way the charge is distributed in it. The radius a is 2.8 fm. Such a charge distribution is necessarily unstable, tending to fly apart; nonelectromagnetic cohesive forces are needed for its stability. This was one reason Dirac argued against an extended electron. A second reason was the existence of the (then) recently discovered neutron (1932). It indicated that not all masses of elementary particles can be derived from electromagnetic self-energy. A third reason was that quantum electrodynamics gave no indication of a finite size, certainly not a size as large as the Abraham-Lorentz size.

The final reason Dirac gave in favor of a point charge is characteristically based on simplicity: as an elementary particle, it should have no internal structure at all. But this demand led inevitably to a divergent Coulomb self-energy. Aiming at a simple equation of motion for this elementary particle, Dirac suggested that this self-energy should be subtracted, omitted from the equation. Thus, he suggested to have no electromagnetic mass. All the mass of the electron is to be of non-electromagnetic origin, m_o,

$$m = m_o + m_{elm} \tag{3}$$

$$m_{elm} = 0 \ . \tag{4}$$

This suggestion seems to have been based on convenience rather than on physical insight; but it was not. Guided by his criteria of beauty and simplicity, he was forced to propose eq. (4) ad hoc.

In the late forties, when Freeman Dyson explained his newly developed renormalization theory to all orders of perturbation expansion to members of the Institute for Advanced Study in Princeton (which at that time was under the directorship of J. Robert Oppenheimer), the following witticism made the rounds: "When a term is infinite, it does not necessarily mean that it is zero because it could also be finite."

Based on that wisdom, one could only combine the (infinite) electromagnetic mass with the (unknown) nonelectromagnetic mass and identify the sum with the observed mass (mass renormalization). Since the observed mass is finite, the nonelectromagnetic mass is thus forced to be infinite also and of opposite sign to the electromagnetic one. (The sensibility of mathematically minded readers may be offended by such talk. In that case the word "infinite" should be replaced by "cut-off dependent").

No such wizardry is suggested by Dirac. He wants a finite nonelectromagnetic mass m_o which is to be identified with the observed mass and he wants no electromagnetic mass. Hence, there is no ugly renormalization suggested in his 1938 paper.

It was not until the mid-seventies that Dirac's conjecture of a vanishing electromagnetic mass was actually proven. The proof was given in an algebraically rather complex but conceptually very simple calculation.[10] The results were later commented on and elaborated.[11, 12]

Nonrelativistic quantum electrodynamics is very much simpler than relativistic quantum electrodynamics, both mathematically and physically. At the same time, it permits considerable insight into certain problems. One of these is the Coulomb self-energy problem because no relativistic effects are involved there.

One considers a (spinless) charged particle of total charge e and nonelectromagnectic mass m_o in interaction with external electromagnetic fields as well as with its own electromagnetic field. The particle is assumed to have at first a finite size (characterized by a "radius" a) and a smooth charge distribution. Even in the nonrelativistic approximation, one cannot ignore the effects of retardation of the solutions to Maxwell's equations. Retardation affects also the interactions between elements of charge within the charged particle (internal retardation).

One can then derive the equation of motion for the position R(t) of the center of the charged particle. One finds

$$m_o \frac{d^2\vec{R}}{dt^2} = \vec{F}_{ext} + \vec{F}_{self} \tag{5}$$

where $\vec{F}_{self}$ is the self interaction force due to the particle's own field. It is an expansion in powers of $(\lambda \frac{d}{dt})$ acting on $d^2\vec{R}/dt^2$,

$$\vec{F}_{self} = - \sum_{n=0}^{\infty} c_n \left(\frac{d}{cdt}\right)^n \frac{d^2\vec{R}}{dt^2} , \tag{6}$$

where $\lambda = \hbar/(m_o c)$ is the Compton wavelength and the coefficients c_n depend on the charge distribution, i.e. on a as well as on λ. The electromagnetic mass m_{elm} is identified with the coefficient c_o,

$$m_{elm} = c_o(a,\lambda). \tag{7}$$

It is some finite positive number depending on the charge distribution.

The great interest in (6) and (7) lies in the fact that these results are quantum mechanically exact, correct to all powers of α (i.e. of $\hbar$). The expansion in (6) is exactly opposite to the one used in relativistic QED where one expands in powers of $\alpha = e^2/\hbar c$, i.e., in <u>reciprocal</u> powers of $\hbar$. The latter expansion cannot have a limit $\hbar \to 0$.

The theory of the <u>classical</u> extended charge is obtained in the limit $\lambda \to 0$ and is finite because of the finite size of the particle. A second limit, $a \to 0$, then leads to the classical point charge with the usual divergent result for m_{elm}.

On the other hand, one can also interchange these two limits; one can first obtain the quantum mechanical point charge ($a \to 0$), and <u>then</u>, from that result, one can obtain the classical point charge ($\lambda \to 0$). The great discovery was that these two limits do not commute: the latter choice of order of limits does not give a divergent m_{elm} but one finds $m_{elm} = 0$ for a classical point charge. This is exactly what Dirac has conjectured, and it proved him to have been correct.

In fact, one finds that already the first limit, $a \to 0$, yields $m_{elm} = 0$: <u>a quantum mechanical point charge has no electromagnetic mass</u>. In order to obtain this result, it was necessary to retain all powers of λ (i.e., of $\hbar$). The coefficients c_n are themselves integrals over series expansions in λ; the series for c_o can, however, be summed into closed form and only then does it yield an integral which vanishes in the point particle limit.

A comparison of the two different orders in which the limits can be taken can be expressed by saying that λ acts like a cut-off in the point limit. But on a deeper level one recognizes that a physically meaningful description can be ensured on the classical level only when the radius of the particle is large compared to the Compton wavelength. A point charge limit is therefore meaningful only on the quantum mechanical level. And that is ensured only when the limits are taken first with $a \to 0$ and then with $\lambda \to 0$.

The result of all this is the conclusion that Dirac's criterion of simplicity and beauty had led him in a seemingly uncanny way to the correct answer.

Parenthetically, one must mention at this point that the beautiful work by Moniz and Sharp[10] also resolved the other two difficulties that beset the Lorentz-Dirac equation: run-away solutions and preacceleration. Both are absent when the classical theory is considered as the classical limit of point charge quantum electrodynamics. But we cannot discuss these problems here.

Unifying Ideas

Some of Dirac's great contributions result from his desire to combine different fields of physics or different mathematical methods into a single unified treatment.

One of his very early contributions arose in the golden days of quantum mechanics. He was seeking an understanding of the relation between Heisenberg's matrix mechanics and the classical canonical formalism of dynamics. What he discovered was the deep correspondence between commutators and Poisson brackets which holds for the algebra of the fundamental dynamical variables.[13] This correspondence has remained the cornerstone of any transition from classical to quantum mechanics, the so-called canonical quantization.

Many years later Dirac again tried to relate the old and proven canonical Hamiltonian dynamics to a relative newcomer, viz. to special relativity. This led to the famous paper on forms of relativistic dynamics[14]: separating the 10 Poincaré generators into those which do not contain explicit reference to the interaction leads to families of hypersurfaces which characterize the dynamical evolution. The three forms of dynamics, the instant form, the point form, and the front form are those in which 6($\vec{P}$ and $\vec{J}$), 6($\vec{K}$ and $\vec{J}$), and 7 of the 10 generators have no explicit reference to the interaction. In Dirac's words these are "specially simple expressions." The desirable simplicity is again emphasized.

In the subsequent two years Dirac published his detailed study of constrained Hamiltonian systems.[15] It answers the problem that arises when the canonical variables are constrained. Such constraints contradict the canonical algebra since that algebra is consistent only for independent variables. Dirac's main goal was to develop this theory so that it can be used to cast the general theory of relativity into a Hamiltonian formulation. Such a formulation is desirable for the purpose of quantization. He succeeded in this effort in 1958.[16] His constraint theory and its application to quantization is beautifully presented in his lectures of 1964 at Yeshiva University.[17]

Dirac's theory of Hamiltonian dynamics with constraints has been amplified and extended by many people and in particular by the Syracuse relativity groups under the direction of Peter Bergmann Its early applications to general relativity have in recent years been widely broadened to applications in various other fields.[18]

The need to introduce more variables than is required by the degrees of freedom of a system occurs in many instances. The associated necessity of constraints then can become a problem when a canonical formulation is demanded. The best known constraint is the Lorentz condition in electrodynamics. But quite generally, any manifestly covariant theory needs more variables than the degrees of freedom require. Thus, one finds application of the constraint theory in relativistic quantum field theory and in particular Yang-Mills theories, so important in present elementary particle physics.

Other applications include the relativistic string and relativistic particle dynamics (both classical and quantum) when formulated without the benefit of a mediating field (direct interaction dynamics). The latter is now receiving a lot of attention because of its potential application to quark and nuclear physics.

Recent developments are easily accessible through useful conference proceedings.[19]

Here one sees again Dirac's ground breaking contributions laying the foundations for far-reaching new developments and techniques.

Pretty Mathematics

"A good deal of my research work in physics has consisted in not setting out to solve some particular problem, but simply examining mathematical quantities of a kind that physicists use and trying to fit them together in an interesting way regardless of any application that the work may have. It is simply a search for pretty mathematics."[20]

This quote is from a recent paper in which Dirac then proceeds to recall how he obtained his famous relativistic wave equation starting out by "playing around with three 2×2 matrices." But he warns that "pretty mathematics by itself is not an adequate reason for nature to have made use of a theory."[20]

Nevertheless, Dirac has repeatedly applied new and pretty mathematics to physics, often mathematics that was not yet established or known by mathematicians and that was developed by them only afterwards. He has perhaps introduced more mathematical innovations into physics than any other theoretical physicist.

One of Dirac's first and best known mathematical innovations is the delta-function.[21] In introducing it he went further than others who had similar ideas.[22] He introduced it

> "...To express in a concise form certain relations which we could, if necessary, rewrite in a form not involving improper functions, but only in a cumbersome way which would tend to obscure the argument."[1]

Thus, the motivation was again simplicity. The lack of a mathematically rigorous foundation was at that time not an issue since most papers on quantum mechanics were purely formal prior to von Neumann's work. A rigorous justification of the delta function was not given until 18 years later in 1945 when Laurent Schwartz, motivated by

Dirac's work, first developed the notion of a distribution. His work started a whole new branch of mathematics.[23]

Related to the use of the delta function is Dirac's use of state vectors that are eigenvectors of operators with continuous spectra (such as the linear momentum). These bra and ket vectors were also introduced on a formal level. They, too, increased the elegance of presentation considerably, as is evident from his Principles of Quantum Mechanics.[1] Their rigorous justification can be found on a rather sophisticated level in a structure that combines distributions and Hilbert space: the rigged Hilbert space.[24] But it can also be justified on the level of conventional Hilbert space.[25]

Orthodox Hilbert space is a space of vectors with positive definite norm. But when unphysical states of negative norm arose, Dirac did not hesitate to advocate a generalization to indefinite norm Hilbert spaces.[3] This permits a formulation of the quantized electromagnetic field in a manifestly covariant gauge (e.g. the Lorentz gauge). Restriction to the orthodox positive definite space would permit only the "ugly" Coulomb gauge formulation (and similar noncovariant gauges). This generalization to indefinite metric Hilbert spaces was later studied by mathematicians in much greater detail.[26]

One of Dirac's brilliant ideas is the notion of magnetic monopoles; their existence would explain charge quantization. It all started in 1931 with a paper[27] that showed the quantization of the product of electric and magnetic charge e and g, $eg/\hbar c = (1/2)n$ (n integer). This same paper is also remarkable because it stated for the first time in Dirac's words that particles and antiparticles must have the same mass according to the Dirac equation (accepting a criticism by J.R. Oppenheimer), and that, therefore, there must exist antiparticles to the negative electrons (found experimentally in the following year by C.D. Anderson) as well as antiparticles to the positive proton (found experimentally in 1955 by O. Chamberlain,

E. Segre, C. Wiegand and T. Ypsilantis).

The theory of magnetic monopoles first deals with non-integrabel phase factors and only in a later development introduces singularity lines or strings.[27] There is one such string attached to each magnetic pole. It ensures the possibility of defining potentials for the electromagnetic fields by making space into a nontrivial manifold. In recent times, these strings have been replaced by a more sophisticated treatment in terms of the differential geometric notions of fiber bundles over nontrivial manifolds. This new development was very strongly motivated by the Dirac monopoles.

These are examples of the mathematical innovations that Dirac introduced so successfully into physics. They also provided fertile grounds for further developments in pure mathematics stimulated by theoretical physics.

It has become clear that Dirac's way of doing physics is not only a science but an art as well. It deals with beauty, simplicity, and generally, with pretty mathematics. It aims for unifying ideas between theories that are apparently in conflict. But underlying it all it uses that rare ability of genius: it senses the right answer.

References

1. P.A.M. Dirac, Principles of Quantum Mechanics, Oxford University Press, Third ed. (1947).
2. P.A.M. Dirac, "The relation between mathematics and physics," James Scott Prize Lecture, February 6, 1939, published in Proc. Roy. Soc. (Edinburgh) 59, 122-9 (1939).
3. P.A.M. Dirac, "The physical interpretation of quantum mechanics," Proc. Roy. Soc. (London) A 180, 1-40 (1942).
4. P.A.M. Dirac, "A new classical theory of the electron, I," Proc. Roy. Soc. (London) A 209, 291-6 (1951).
5. P.A.M. Dirac, "A new classical theory of electrons, II," Proc. Roy. Soc. (London) A 212, 330-9 (1952).
6. P.A.M. Dirac, "An extensible model of the electron," Proc.

Roy. Soc. (London) A 167, 148 (1938).

7. P.A.M. Dirac, "A positive energy relativistic wave equation," Proc. Roy. Soc. (London) A 322, 435-45 (1971).

8. F. Rohrlich, "The Electron: Development of the First Elementary Particles Theory" in J. Mehra (ed.), The Physicist's Conception of Nature, p. 331-369, Reidel Publishing Co., Dordrecht-Holland, 1973.

9. P.A.M. Dirac, "Classical theory of radiating electrons," Proc. Roy. Soc. (London) A 167, 148 (1938).

10. E.J. Moniz and D.H. Sharp, Phys. Rev. D 10, 1133 (1974) and 15, 2850 (1977).

11. F. Rohrlich, Acta Phys. Austriaca 44, 375 (1975).

12. H. Grotch, E. Kazes, F. Rohrlich, and D.H. Sharp, Acta Phys. Austriaca 54, 31 (1982).

13. P.A.M. Dirac, "The fundamental equations of quantum mechanics," Proc. Roy. Soc. (London) A 109, 642-53 (1925).

14. P.A.M. Dirac, "Forms of Relativistic Dynamics," Rev. Mod. Phys. 21, 393-9 (1949).

15. P.A.M. Dirac, "Generalized Hamiltonian dynamics," Canad. J. Math. 2, 129-48.

16. P.A.M. Dirac, "The theory of gravitation in Hamiltonian form," Proc. Roy. Soc. (London) A 246, 333-43 (1958).

17. P.A.M. Dirac, Lectures on Quantum Mechanics, Yeshiva University, N.Y. 1964.

18. A fine collection of references up to about 1979 can be found in the recent survey by K. Sundermeyer, Constrained Dynamics, Lecture Notes in Physics 169, Springer-Verlag, Berlin 1982.

19. J. Llosa, Relativistic Action at a Distance: Classical and Quantum Aspects Lecture Notes in Physics 162, Springer-Verlag, Berlin 1982.

20. P.A.M. Dirac, "Pretty Mathematics," Int'l J. Theor. Phys. 21, 603-5 (1982).

21. P.A.M. Dirac, "The physical interpretation of quantum mechanics,"

Proc. Roy. Soc. (London) A 113, 621-41 (127).

22. G. Kirchhoff, Berliner Ber. p. 641 (1882); O. Heaviside, Proc. Roy. Soc. (London) A 52, 504 (1893) and 54, 105 (1893).

23. L. Schwartz, Theorie des Distributions, Paris, 1957.

24. This subject is made easily accessible to physicists by A. Bohm, The Rigged Hilbert Space and Quantum Mechanics Lecture Notes in Physics 78, Springer-Verlag, Berlin 1978.

25. J.M. Jauch in Aspects of Quantum Theory (A. Salam and E.P. Winger, eds.) Cambridge Univ. Press 1972, p. 137.

26. K.L. Nagy, State Vector Spaces with Indefinite Metric in Quantum Field Theory, Akademiai Kiadó, Budapest, 1966.

27. P.A.M. Dirac, "Quantized singularities in the electromagnetic field," Proc. Roy. Soc. (London) A 133, 60-72 (1931).

28. P.A.M. Dirac, "The theory of magnetic poles," Phys. Rev. 74, 817 (1948).

A SEARCH FOR PROTON DECAY INTO $e^+\pi^\circ$

IRVINE-MICHIGAN-BROOKHAVEN COLLABORATION

R.M. Bionta,[2] G. Blewitt,[4] C.B. Bratton,[5] B.G. Cortez,[2,a] S. Errede,[2] G.W. Foster,[2,a] W. Gajewski,[1] M. Goldhaber,[3] J. Greenberg,[2] T.J. Haines,[1] T.W. Jones,[2 7] D. Kielczewska,[1,b] W.R. Kropp,[1] J.G. Learned,[6] E. Lehmann,[4] J.M. LoSecco,[4] P.V. Ramana Murthy,[1,2,c] H.S. Park,[2] F. Reines,[1] J. Schultz,[1] E. Shumard,[2] D. Sinclair,[2] D.W. Smith,[1,d] H.W. Sobel,[1] J.L. Stone,[2] L.R. Sulak,[2] R. Svoboda,[6] J.C. van der Velde,[2] and C. Wuest.[1]

(1) The University of California at Irvine, Irvine California 92717

(2) The University of Michigan, Ann Arbor, Michigan, 48109

(3) Brookhaven National Labratory, Upton, New York 11973

(4) California Institute of Technology, Pasadena, California 91125

(5) Cleveland State University, Cleveland, Ohio 44115

(6) The University of Hawaii, Honolulu, Hawaii 96822

(7) University College, London, U.K.

(presented by F. Reines)

Observations were made 1570 mwe underground with an 8000 metric ton water Cherenkov detector. During a live-time of 80 days no events consistent with the decay $p \to e^+\pi^\circ$ were found in a fiducial mass of 3300 metric tons. We conclude that the limit on the lifetime for bound plus free protons divided by the $e^+\pi^\circ$ branching ratio is

$\tau/B > 1.9 \times 10^{31}$ yr. (90% confidence). Observed cosmic ray muons and neutrinos are compatible with expectations.

I. INTRODUCTION

We have built and operated a large water Cherenkov detector deep underground in order to search for nucleon decay. We have chosen as the initial goal of our experiment to look for proton decay to the final state e^+ and π°. This two body mode gives a clear back to back decay signature which is especially well-defined and distinct from background in a water Cherenkov detector.

The simplest grand unified gauge theory, minimal SU(5), predicts a partial lifetime $\tau/B=4.5\times10^{29\pm1.7}$ years for the $p \to e^+\pi^\circ$ decay mode.[1] Therefore, in our detector, after 80 days of exposure (2.4×10^{32} proton years) using $\frac{\tau}{B} = 2.2\times10^{31}$ years, we expect to observe at least seven such events.

Previous searches for nucleon decay in iron at the Kolar Gold Fields[2] and the Mont Blanc Tunnel[3] have reported a few candidate events tentatively ascribed to various modes of proton decay.

II. DETECTOR

The Irvine-Michigan-Brookhaven (IMB) detector[4] is located at a depth of 1570 meters of water equivalent in the Morton-Thiokol Salt Mine east of Cleveland, Ohio. It consists of a large rectangular volume of purified water (17m×18m×23m) viewed from its six faces by 2048 photomultiplier tubes (PMT), each of 12.5 cm diameter, located on a rectangular grid of ~ 1m spacing. The total sensitive volume of the detector is 8000 metric tons (tonnes) and the fiducial volume, inset by two meters from the planes of the PMT's is 3300 tonnes.

A relativistic charged particle traversing the detector produces a cone of Cherenkov light with a half opening angle of 41° relative to its direction of motion. The particle continues to emit Cherenkov light until its velocity falls below 0.75C. The time of photon arrival and pulse height are recorded independently by each PMT, providing information which allows the reconstruction of position, direction and energy of particles moving in the detector.

Two distinguishing characteristics of this detector are its large sensitive mass and its ability to determine unambiguously the the sense of track direction. The uniform sensitivity of the active medium makes the energy resolution of the detector relatively independent of the fluctuations of electromagnetic shower development and Fermi motion of the decaying proton.

The resolution and absolute energy calibration of the detector in time, position, direction and energy is evaluated by the use of cosmic ray muons and programmable light sources in the detector volume. The light sources have variable position, intensity, and firing time. They are also used to determine the attenuation of length (>30m) for light in the wavelength region of interest (300-450nm). Our system of PMT's and electronics is sensitive to light at the single photoelectron level.

The trigger threshold corresponds to the light output from a 50 MeV electron in the detector.[5] PMT's firing up to 7.5 μsec after the initial trigger are also recorded providing identification of the $\mu \rightarrow e\nu$ decay sequence with 60% efficiency.

On the basis of the measured efficiency of PMT's and the geometry of our detector, computer event simulation predicts a signal from 170 ± 18 (statistical) ± 18 (systematic) tubes if a proton decays to $e^+\pi^\circ$ inside the fiducial volume. Event simulation programs were calibrated using data from single cosmic ray muons.

The time resolution of PMT's at the single photoelectron level (11 ns FWHM) is the primary factor in determining the accuracy of the vertex position for $p \rightarrow e^+\pi^\circ$. For this mode resolution is typically ±60 cm, which has been checked using point sources of light at various positions in the detector.

For two-track events with an opening angle >100° and where each track gives rise to signals from at least 40 PMT's, the cones from both tracks are distinguishable and the opening angle has a typical uncertainty of ±15°. For $p \rightarrow e^+\pi^\circ$ the two photons from π° decay are generally not distinguishable, but the π^0 shower gives a cone similar to that from the e^+. The detector was checked by measuring the angular-distribution, stopping muon rate and muon life time,(4) which are consistent with accepted values.

III. ANALYSIS

Reduction of the 2.3×10^5 triggers/day (primarily due to through-going muons) is accomplished with three independent and complementary analyses. In the first stages of analysis, cosmic ray muons are rejected by various energy and topological cuts. The times and spatial locations of the lit tubes are then used to find the vertex, assuming that the light originates from a point source. A fiducial cut two meters in from the planes of the tubes rejects entering tracks by more than a factor of 300.[6] Different pattern recognition methods, including interactive scanning by physicists, are applied to the surviving sample. Events are fit to one and two track hypotheses and track directions are determined using the known properties of Cherenkov light. The goal of the filtering procedures is to save all events which originate inside the fiducial volume.

The survival efficiency of $p \rightarrow e^+\pi^\circ$ events through three procedures is estimated to be 0.9±0.1 based on computer simulations whose validity has been verified using cosmic ray muons and artificial light sources. Cross correlation between the three analyses on data events which have properties similar to those expected for $p \rightarrow e^+\pi^\circ$ further confirm the estimated detection efficiency.

IV. SUMMARY OF RESULTS

With 2.4×10^{32} proton years of exposure, we find 69 interactions in the fiducial volume which light at least 45 PMT's. Within the statistical accuracy of the sample, the above events are uniform in vertex position and isotropic in track direction. There is no evidence for contamination of the sample by entering tracks or from events generated by the predominantly downward-going muon associated background (Figure 1). The fraction of fully-contained events

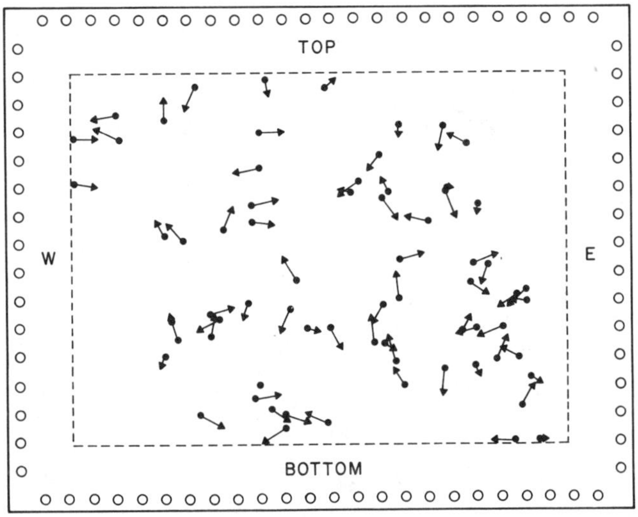

FIG. 1. Side view of the detector indicating vertex positions for one and two-track events. Arrow indicates the projection of the direction unit vector for the best fit tracks in this view. PMT positions are indicated near the outside (solid line) of the water volume. The dashed line represents the boundary of the software controlled fiducial volume.

exhibiting $\mu \rightarrow e$ decay is 0.4±0.1 after correcting for our detection efficiency of 0.6±0.1.[7] The distribution of visible energy, E_{min}, for the events is displayed in Figure 2. The curve in this Figure was calculated by passing simulated neutrino events through our filtering cuts. The neutrino events were generated using data from a wide band neutrino exposure in the Gargamelle bubble chamber.[8] The spectrum was adjusted to agree with that expected for atmospheric neutrinos and corrections were made for the lack of electron neutrinos in the Gargamelle data. We calculate the absolute rate of neutrino interactions which pass our energy cuts to be 0.9±0.4 per day, based on estimated fluxes[9] and cross sections[10] appropriate to the energy range we observe. Our estimated efficiency for the detection of atmospheric neutrinos is 0.7±0.2. The predicted rate could be in error by as much as a factor of two due to uncertainties

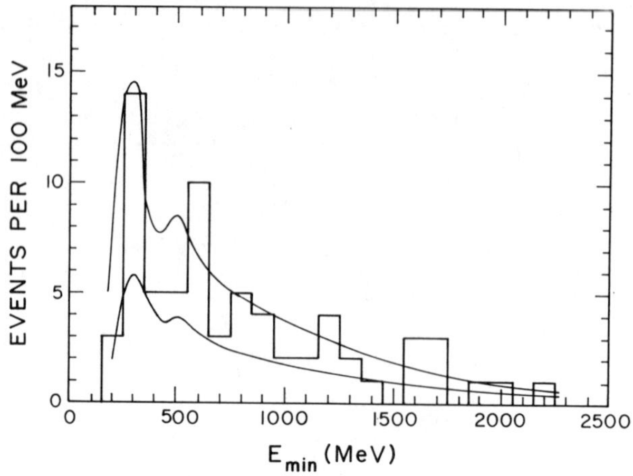

FIG 2. Energy (E_{min}) distribution of the 69 contained events. The expected distribution from atmospheric neutrinos lies between the upper and lower curves. The uncertainty arises primarily from the atmospheric flux estimate. The secondary peak near 500 MeV is due to events with observed muon decays whose visible energy has accordingly been increased by 230 MeV.

arising from the calculation of the neutrino flux. The results are thus seen to be consistent with expectations.

The 69 contained events can be summarized as follows:

1) Sixty-six single or multi-track events which do not possess a track lighting more than 40 tubes in the backward hemisphere, and hence are outside the angle and energy requirements for $p \rightarrow e^+\pi^\circ$.

2) Two wide angle two-track events, both of which have a muon decay signal and hence do not qualify as $p \rightarrow e^+\pi^\circ$.[11]

3) One two-track event in which 340 PMT's were lit, about a factor of two greater than expected for $p \rightarrow e^+\pi^\circ$ in our detector. In addition the opening angle 115°±15° was outside the predicted range (>140°) for free or bound proton decay to $e^+\pi^\circ$. The characteristics of these two-track events are summarized in Table I.

TABLE 1. WIDE ANGLE TWO-TRACK EVENTS

Event Number	Number of PMT's Lit	Minimum Energy E_{min} in MeV (a)	Track Opening Angle Deg.	Obs. Muon Decays
151-35037	188	1230±135	135±7	1
225-7794	166	1180±120	125±25	1
388-19376	340	1700±170	115±15	0

(a) Does not include possible systematic errors of 15%. E_{min} is the minimum energy of the event as inferred from the Cherenkov light yield after making a 230 MeV correction for rest mass and Cherenkov yield of observed muons.

Results on the other nucleon decay modes and a more complete analysis of neutrino interactions will be discussed in future publications.

V. CONCLUSIONS

For free protons, we obtain a lifetime limit independent of nuclear effects. Since 1/5 of the protons in our detector are not bound in nuclei, we can conclude that the partial lifetime for free protons is $\frac{\tau}{B} > 1.9\times10^{31}$ years.

For free plus bound protons we assume a 40% loss of events in the oxygen nucleus and ignore possible lifetime changes from nuclear effects.[12] Using 2.3 events to obtain a 90% confidence limit, we calculate τ/B as follows:

$$\tau/B > Nf\varepsilon_n \varepsilon\tau/2.3 = 6.5 \times 10^{31} \text{ years}$$

Where:

$N = 2.0 \times 10^{33}$ nucleons

$f = 10/18$ proton fraction in water

$\varepsilon_n = 0.68$ average nuclear $\pi°$

survival efficiency (oxygen plus free protons)

$\varepsilon = 0.9\pm0.1$ detection efficiency

$T = 80/365$ years

Acknowledgements

We gratefully acknowledge the efforts of the employees of Morton-Thiokol, Inc. who operate the Fairport mine, especially the contributions of managers J. Davis and R. Lark and engineers B. Cummings and B. Lewis.

We wish also to thank C. Cory, E. Hazen and the many other people whose help was invaluable in making this detector a reality.

REFERENCES AND FOOTNOTES

a) Also at Harvard University;
Permanent addresses:
b) Warsaw University, Poland
c) Tata Institute of Fundamental Research, Bombay, India;
d) Now at University of California, Riverside, 92521

1. H. Georgi and S.L. Glashow, Phys. Rev. Lett. 32, 438 (1974); P. Langacker, Phys. Rev. 72, 185 (1981) and Proceedings of the 1982 Workshop on Proton Decay, ANL-HEP-82-24 (D.S. Ayres, Ed.) P. 64; M.A.B. Beg and A. Sirlin, Phys. Rep. 88, 1 (1982). See also N. Isgur and M.B. Wise, Phys. Lett. 117B, 179 (1982), and W.J. Marciano, BML 31036, presented at Orbis Scientiae 1982.
2. M.R. Krishnaswamy, et. al., Phys. Lett. 115B, 4, p. 349, (1982).
3. G. Battistoni, Phys. Lett. 118B, p. 461 (1982).
4. R. Bionta, et. al., Proceedings of the 1982 Moriond Conf., Tran Than Van, ed., (1982).
5. More precisely, the trigger requires that either > 12 PMT's fire within 50 nsec. or that > 3 PMT's fire in any 2 of 32 groupings of 8 × 8 PMT's in 150 nsec.
6. The actual fiducial volume cut is topology dependent and can be as close as 1.0 m from the tube planes in rare cases.

This work is supported in part by the U.S. Department of Energy.

Our nominal "fiducial mass" is obtained by determining the fraction of simulated events generated throughout the entire detector which pass all of our filtering cuts.

7. In the energy range with which we are concerned, the ν_μ/ν_e ratio is expected to be ~ 2/1. Due to the different sensitivities of our detector to electrons and muons, with the requirement of > 45 PMT's we expect that the observed ratio will be ~ 0.5.
8. H. Deden, et. al., Nuclear Physics B85, 269 (1975).
9. J.L. Osborne and E.C.M. Young in "Cosmic Rays at Ground Level" (A. Wolfendale, ed.) 1973.
10. S.J. Barish, et. al, Phys. Rev. D. 19, 2521 (1979).
11. These two events have 15 and 12 PMT's respectively associated with a muon decay electron in a time window of 100 ns. Both of these numbers are far above the threshold of 6 PMT's set by accidental coincidences in our second timing scale.
12. C.B. Dover, M. Goldhaber, P.L. Trueman and L.L. Chau, Phys. Rev. D 24, 2886 (1981).

DIVERSITY IN HIGH ENERGY PHYSICS

A.D. Krisch

Randall Laboratory of Physics
The University of Michigan
Ann Arbor, Michigan 48109

I will start the session with a short talk about the need for diversity in high energy physics. I consider the proton decay experiment we just heard about an excellent ecample of the value of diversity.

This round table has some of the world's most distinquished experts on particle accelerators, which are our most valuable tool for studying the inner structure of elementary particles. I had hoped to have lectures about each different type of high energy facility developed in this century:

Proton Accelerators	$\bar{p}$ - p Colliders
Electron Accelerators	$e^+ - e^-$ Colliders
p - p Colliders	e - p Colliders

I just learned that Professor MacDaniel had to cancel his lecture on "High Luminosity $e^+ - e^-$ Facilities"; thus, I want to encourage all participants to comment on $e^+ - e^-$ colliders. Fortunately, this area is not totally unrepresented, since we have on the panel the director of DESY and a man with some possible relation to Cornell's R. R. Wilson Laboratory

I want to stress the vital need for a wide variety of high energy facilities, because recently some of us may have forgotten this need in our enthusiasm for our own speciality. From time to time, some people express the firm belief that only one type of particle should be studied. Others tell us, with absolute certainty, that nothing interesting will occur in the"desert" below 10^{16} GeV.

In my opinion, such firm and absolute beliefs about still unmeasured phenomena are more appropriate to theological studies. We scientists must always remember that there are many very important areas of human activity which cannot be studied by the scientific method with its rigid criteria of experimental proof. Nevertheless, I hope that we all agree that these rigid scientific criteria should be applied to elementary particle studies.

Thus, until we have conclusive experimental evidence pointing in one direction, I firmly believe that we should explore as many different research frontiers as possible, within the budgets allocated by our governments. It seems vital that we have e facilities, p facilities, e - p facilities, and decay facilities on this planet. I also feel that we should stress with equal weight the various frontier capabilities:

High Energy	Polarization
High Luminosity	and possibly others unknown to me

With our present knowledge of strong interactions, I do not understand how any responsible physicist could conclude that we should no longer study one of these areas. I give an infinitely higher priority to maximizing the number of different ways we study elementary particles, rather than maximizing the speed with which we study one special area. We should recall how quickly our colleagues in the 1890's convinced themselves that they understood everything about physics. They were, of course, correct except for quantum mechanics and relativity.

I hope that we learned something from the 1890's and do not prematurely decide that "QCD" or "the standard model" or "Regge Poles" are scientific facts. High energy physics may be the most difficult and abstract of scientific study; and real successes are rare. We should, therefore, be especially careful not to lower our standards to accept a success or belief that is not rigorously justified. By maximizing the number of <u>independent</u> high energy physics capabilities, in theory, in experiment, and in accelerators, we have <u>some</u> insurance against accepting a false success.

The dissertators today will be Volker Soergel discussing "High Energy e - p Facilities", Robert Wilson discussing "Very High Energy p - p Facilities" and Nicholas Samios discussing "High Luminosity p - p Facilities". Our annotators, Ernest Courant and Paul Reardon, will comment on these. I want to encourage strongly all participants to make comments or suggestions. Suggesting a new technique for studying the laws of physics would be an excellent way to honor our most distinguished colleague, Professor Dirac.

SOME REMARKS ON PERFORMANCE OF VERY HIGH ENERGY COLLIDERS*

E. D. Courant

Brookhaven National Labratory

Upton, New York 11973 and

State University of New York at Stony Brook

Stony Brook, New York 11794

We have heard from R. R. Wilson of the exciting possibility of a huge "desert machine" with 20 TeV protons and antiprotons, and N. Samios has stressed that high luminosity is desirable, especially at very high energies.

I want to discuss how the luminosity in a very large p-$\bar{p}$ storage ring scales with the magnetic field of the ring magnets, among other parameters.

One of the limitations on performance is given by the beam-beam tune shift. Experience at the SPS seems to indicate that, with colliding bunched p and $\bar{p}$ beams, this tune shift should not exceed 0.003 per crossing.

With round beams colliding head on, the tune shift equals

$$\Delta\nu = \frac{Rr_p I/e}{2kc\varepsilon_t} = \frac{Nr_p}{4\pi\varepsilon_t}$$

*Work performed under the auspices of the U. S. Dept. of Energy

and the luminosity is

$$L=\frac{\gamma(I/e)^2 R}{2k\varepsilon_t\beta^*}=\frac{2k\varepsilon_t c\gamma}{r_p^2 R\beta^*}(\Delta\nu)^2 ,$$

where ε_t is the invariant transverse beam emittance at the rms oscillation amplitude, β^* the amplitude function at the crossing point, k the number of bunches in each beam, each containing N particles, and r_p is the proton radius.

Keeping ε_t and k constant, we see that the current for constant $\Delta\nu$ scales as 1/R, i.e. the number of particles per bunch is constant, while the luminosity scales with 1/R and $1/\beta^*$.

How does β^* scale? The low-beta crossing point is obtained by focusing the beam to a small size with a quadrupole doublet (or triplet or quadruplet) near the crossing point. For comparable machine lattices, β scales with the focal length of the quadrupoles. If magnet apertures are fixed, and quadrupoles are assumed to have pole fields proportional to the magnetic field of the ring magnets, then the focal length, and hence β^*, scales as $R^{1/2}$ (the tune ν then also scales as $R^{1/2}$), and the luminosity, scales as $R^{-3/2}$, i.e. stronger field and smaller radius gives enhanced performance.

Numerically, what do we get?

Assume the invariant emittance is blown up to 10 mm-mr (as compared to about 2.5 in most existing machines). With $\Delta\nu$=0.003 we then get 2.3×10^{11} particles per bunch. With eight crossing regions (k=4) and β^* scaling from 2m for R=.5 km. we find:

B	2T	5T	10T
R	50 km	20 km	10 km
β^*	22 m	14 m	10 m
L*	1.8×10^{29}	6.9×10^{29}	1.9×10^{30}

Better designed parameters can undoubtedly be found, and the scaling laws, especially for β^*, may well be modified. But it is clear that the performance capability of very high energy storage ring colliders is very much enhanced by building them with the strongest feasible magnetic fields and smallest circumferences.

GLUEBALLS

Sydney Meshkov*

University of California

Los Angeles, California 90024

Abstract

The current status of various glueball properties such as level ordering and masses is reviewed. Current glueball candidates ι(1440), θ(1670), and ϕϕ enhancements at 2160 and 2320 MeV are examined. A simple model which incorporates the mixing of the glueball candidate ι(1440) with quarkonium states η(549) and η'(958), and of the θ(1670) with f(1270) and f'(1515) is presented. Neither the ι(1440) nor the θ(1670) can be consistently interpreted as a glueball in this framework. The leading glueball candidates currently are the ϕϕ enhancements.

Glueballs[1,2(a-f)]--colorless, flavorless composites of two or more gluons--should exist, according to the gospels of quantum chromodynamics. In this brief review I summarize the current (~January, 1983) status of theory and experiment for various aspects of glueball spectroscopy. What we know currently about glueball candidates ι(1440), θ(1670) and ϕϕ enhancements at 2160 and 2320 MeV is discussed. In order to help analyze the properties of the ι(1440)

---

*On leave from National Bureau of Standards, Washington, D.C. 20234

and θ(1670), I present a simple model[3] which emphasizes the role of the glueball-quarkonium mixing. The ι(1440) is mixed with η(549) and η'(958). The θ(1670) is mixed with f(1270) and f'(1270). In this mixing framework neither state can be consistently interpreted as a glueball, so I conclude that ι(1440) and θ(1670) are most likely not glueballs. The leading glueball candidates currently are the φφ enhancements.

I. Spectroscopy

(a) Ordering.

The simplest models which give the order and J^{PC} of glueball states[4(a, b, c)] arise from potentials for confinement; no masses are computed in this approach. This has been presented in detail before.[2(e)] We mention the results briefly. The spectrum is divided into 2 and 3 gluon sectors and is shown below in Fig. 1.

$$
\begin{array}{ll}
\underline{\quad}\ \boxed{0^{--}} & \begin{array}{l}(1s)\ (1p)^2 \\ (1s)^2\ (1d) \\ (1s)^2\ (2s)\end{array} \\[1em]
& 0^{++}\ \ldots\ 3^{++} \\
\underline{\quad}\ \boxed{0^{+-}},\ 1^{+-},\ \boxed{2^{+-}},\ 3^{+-} & (1s)^2\ (1p) \\
\underline{\quad}\ 2^{++},\ 0^{++}\ \ldots\ 4^{++}\qquad \underline{\quad}\ 0^{-+};\ 1^{--},\ 3^{--} & (1s)^3 \\
\underline{\quad}\ 0^{-+},\ \boxed{1^{-+}},\ 2^{-+} & \\
\underline{\quad}\ 0^{++},\ 2^{++} & \\
\qquad\qquad 2g & \qquad\qquad 3g
\end{array}
$$

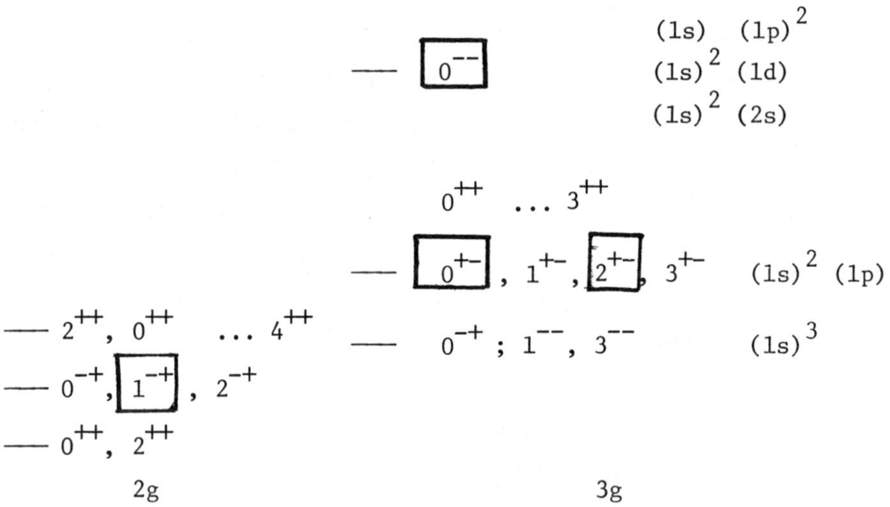

Fig. 1. 2g and 3g levels.

Note that an oddball, 1^{-+}, occurs for the first time at the L = 1 level. Oddballs[4,5] are glueballs with J^{PC} quantum numbers that do not occur in $\bar{q}q; L$ mesons, i.e.,

0^{--}; 1^{-+}, 3^{-+}, ...; 0^{+-}, 2^{+-}, ...

(b) Masses.

1. Lattice gauge theories.

The most active field for estimating glueball masses at present is that of Lattice Gauge Theories (LGT). I list below (fig. 2) three estimates of glueball spectra[5(a,b,c)] expressed as multiplets of $m(0^{++})$. These are representative of what is going on in the field now.

Berg[5(a)] Billoire		Ishikawa[5(b,c)] Schierholz Teper (and Sato)	
1^{-+}	4.8		
2^{--}	4.8		
0^{--}	4.6		
1^{+-}	4.2		
3^{+-}	3.7	0^{--}	3.8
2^{-+}	3.5	1^{-+}	2.3
3^{++}	3.1	2^{++}	2.3 (3.3)
2^{++}	2.5	0^{-+}	2.0
0^{++}	1	0^{++}	1

Fig. 2. Lattice Gauge Theory Masses (Multiples of $m(0^{++})$).

Clearly, a major problem is the determination of $m(0^{++})$. If it is taken at $\sim$ 1 GeV as estimated earlier by Berg, Billoire and Rebbi,[6] then the excited states lie primarily in the 2-5 GeV range. These differ markedly from the earlier estimates of 1-2 GeV range. On the other hand, if we use the preferred value of IST,[5] i.e. $m(0^{++})$ = 0.72, then $m(0^{-+})$ = 1.43 GeV and $m(2^{++})$ = 1.65 GeV, corresponding to $\iota(1440)$ and $\theta(1670)$ respectively. Note that in all three calculations, the oddballs lie high. Overall, the three efforts are fairly consistent.

2. Massive gluons.

Cornwall and Soni[7] consider glueballs to be bound states of massive gluons. They describe the gluon dynamics as massive 1^{--} fields interacting through a breakable string. Their string potential is

$$V_S(r) = 2m\,(1 - e^{-r/r_o}) \quad ,$$

where m is an effective gluon mass, determined by C. Bernard[8] to be ≃ 500 MeV; they take $r_o = 0.6f$. For these parameters, they fit the ι(1440) and θ(1670) masses quite well. Were they to use an effective gluon mass of 750 MeV as determined earlier by Parisi and Petronzio,[9] then all of their glueball masses must be raised by 500 MeV. Spin dependence in this model comes from one gluon exchange.

3. Bag model.

Several bag model calculations with somewhat different characteristics have been carried out. The most recent one of Carlson, Hansson and Peterson[10] has an 800 MeV 0^{-+} ground state and high lying (∿ 1500 MeV) 0^{++} and (∿ 1900 MeV) 2^{++} excited states. The earlier calculations[11] give a more usual two gluon spectrum with degenerate 0^{++} and 2^{++} ground states at 960 MeV, but give a 3g spectrum with parity opposite to that of the potential model approach.[11,2(e)]

The general conclusions that I draw about masses are:

1. $m(J^{PC})/m(0^{++})$ are getting pinned down by LGT;

2. We really don't know where $m(0^{++})$ is nor is it likely to be worthwhile to look for it. In addition to theoretical problems associated with mixing with the vacuum, anyone who has done ordinary meson spectroscopy knows how hard it is to find and analyze 0^{++} states. A notable exception is the χ(3410) ^3P_o $c\bar{c}$ state. We should concentrate on looking for various excited glueball states;

3. Excited glueball masses may very well be higher than formerly believed, i.e. ≥ 2 GeV.

II. Experimental Glueball Candidates [2(f)]

1. What has by now become a "classic" method for finding glueballs is to look for them in the radiative decay of a vector meson (ψ or Υ),[2(d)] i.e. ψ (or Υ) $\to \gamma G$. The estimate

$$\frac{\Gamma\ (\psi\to\gamma gg)}{\Gamma\ (\psi\to ggg)} = \frac{16\alpha}{5\alpha_s}$$

implies that $B(\psi\to\gamma G) \simeq 10\%$.

(a) $\iota(1440)$ $J^{PC} = 0^{-+}$

This was the first candidate and has remained a highly controversial resonance. It was first observed by the Mark II detector in the decay mode

$$\psi \to \gamma\ K_S\ K^{\pm}\pi^{\mp} \quad .$$

It was subsequently observed by the Crystal Ball in the decay

$$\psi \to \gamma\ K^+K^-\ \pi^o \quad .$$

The $\iota(1440)$ seems to decay primarily into $\delta\pi$ where $\delta \to K^+K^-$. $\delta(980)$ is an $I = 1, J^{PC} = 0^{++}$ resonance. The parameters of the $\iota(1440)$ are shown in Fig. 3, (from Reference 2(b)).

One of the few relatively new pieces of data is the observation[12]

$$B(J/\psi \to \gamma\iota)\ B(\iota \to \eta\pi\pi) < 2 \times 10^{-3}\ (90\%\ CL)$$

$$\Longrightarrow \frac{B(\iota \to \eta\pi\pi)}{B(\iota \to K\bar{K}\pi)} \leq \frac{1}{2}\ .$$

The $\iota(1440)$ does not decay to $\eta\pi\pi$ at a statistically significant level, yet it should. There have been many reviews discussing whether or not $\iota(1440)$ is a glueball, a $q\bar{q}$ radial excitation, a mixture, $q\bar{q}g$, $q\bar{q}q\bar{q}$, etc. I address this question in Section III.

$\iota(1440)$ Parameters

Parameter	Experimental Measurement Mark II	Crystal Ball
M (MeV)	$1440 \begin{smallmatrix} +\ 10 \\ -\ 15 \end{smallmatrix}$	$1440 \begin{smallmatrix} +\ 20 \\ -\ 15 \end{smallmatrix}$
Γ (MeV)	$50 \begin{smallmatrix} +\ 30 \\ -\ 20 \end{smallmatrix}$	$55 \begin{smallmatrix} +\ 20 \\ -\ 30 \end{smallmatrix}$
$B(\psi \to \gamma\iota) \times B(\iota \to K\bar{K}\pi)$[a]	$(4.3 \pm 1.7) \times 10^{-3}$[b]	$(4.0 \pm 1.2) \times 10^{-3}$
C	+	+
J^P		0^-

[a] I = 0 was assumed in the isospin correction.

[b] This product branching ration has been corrected by Scharre to account for the efficiency correction required under the spin 0 hypothesis.

Fig. 3. From Ref. 2(b).

(b). $\theta(1670)$ $J^{PC} = 2^{++}$

The second SPEAR candidate, the $\theta(1670)$, was originally observed by the Crystal Ball in the decay $\psi \to \gamma\eta\eta$. By analyzing the $\eta\eta$ yield, and not putting in a term to separate the $f' \to \eta\eta$ contribution, it was found that $M(\theta) = 1640 \pm 50$ MeV; $\Gamma(\theta) = 220 \begin{smallmatrix} +\ 100 \\ -\ 70 \end{smallmatrix}$ MeV; $B(\psi \to \gamma\theta)\ B(\theta \to \eta\eta) = (4.9 \pm 1.4 \pm 1.0) \times 10^{-4}$ and that $\theta(1640) \not\to \pi\pi$.

Since then, a variety of final states X in the 1650-1710 MeV region have been found in the decay $\psi \to \gamma X$. A recent Crystal Ball analysis[2(f),13] makes a two resonance fit to the $\eta\eta$ channel and finds $M(\theta) = 1670 \pm 56$ MeV. In Fig. 4, we list a summary of decay modes, branching fractions and masses in the θ region. A Mark II two-resonance fit[14] of the K^+K^- channel gives $M(\theta) = 1700 \pm 20$ MeV;

a measurement in the $\rho^\circ\rho^\circ$ channel[15] is fit with a single resonance at 1650 ± 50 MeV and a measurement in the $\eta\pi\pi$ channel gives 1710 ± 45 MeV.[12] Of the results given above, the two lowest mass values come from experiments which have lower statistics than the others. It seems prudent to take a somewhat higher mass for the θ than 1640 MeV; we use M(θ) = 1670 ± 50 MeV in subsequent calculations.

θ(1670) 2^{++} 2(f),12,13,14,15

Comments	Decay Mode	Mass (MeV)	Γ(MeV)	B(J/ψ→γθ) B(θ→)
no f'	$\eta\eta$	1640 ± 50	220^{+100}_{-70}	$(4.9\pm1.4\pm1.0)\times10^{-4}$
with f'	$\eta\eta$	1670 ± 50	160 ± 80	$(3.8\pm1.6)\times10^{-4}$
	$\pi\pi$			$< 2.4 \times 10^{-4}$
with f'	$K\bar{K}$	1700 ± 20	156 ± 30	$(12.4\pm1.8\pm5.0)\times10^{-4}$
	$\rho^\circ\rho^\circ$	1650 ± 50	200 ± 100	$(1.25\pm.35\pm.4)\times10^{-4}$
	$(M_{\rho\rho} < 2.0$ GeV)			
	$\eta\pi^+\pi^-$	1710 ± 45	530 ± 110	$(3.5\pm.2\pm.7)\times10^{-3}$
	$\eta\pi^\circ\pi^\circ$	1710 ± 45	530 ± 110	$(2.3\pm.3\pm.8)\times10^{-3}$
	$f' \to K\bar{K}$			$(1.6\pm.5\pm.8)\pm10^{-4}$
	$f' \to \eta\eta$			$(.9\pm.9)\times10^{-4}$

$$\frac{\Gamma(\theta\to\pi\pi)}{\Gamma(\theta\to\eta\eta)} < 1.1$$

Fig. 4. Decays in the θ region.

(c) *L*(2160) and *L*(2320) $J^{PC} = 2^{++}$

A BNL/CCNY collaboration in the reaction $\pi^- p \to K^+K^-K^+K^- n$

22.6 GeV/c sees evidence for the reaction $\pi^- p \rightarrow \phi\phi n$ in the region 2100-2400 MeV.[16] A detailed analysis shows that this peak is fitted with two 2^{++} $\phi\phi$ resonances. The low resonance has a mass of 2160 ± 50 MeV, Γ = 310 ± 70 MeV and is s-wave; the high one has M = 2320 ± 40 MeV, Γ = 220 ± 70 MeV and is d-wave.

The reaction violates the Zweig rule badly, yet the cross section is comparable to that for Zweig allowed processes. Lindenbaum[16] speculates that these enhancements came from the decay of glueballs.

The existence of two peaks in the same channel at 16 GeV/c is seen by the ABBCGLM Collaboration who find two interfering 2^{++} states in $\phi\phi$.[17] The BCGMSP Collaboration in $K^- p \rightarrow \phi\phi\Lambda$ at 8.25 GeV/c finds a $\phi\phi$ enhancement in the region discussed above.[18]

III. Mixing Model of Fishbane, Karl and Meshkov [19]

We examine the consequences of using a simple glueball-quarkonium mixing model in an attempt to see whether the $\iota(1440)$ and $\theta(1670)$ can be described as glueballs. The mass matrix that we use was introduced by Fuchs,[20] de Rujula, Georgi and Glashow[21] and Isgur.[22] It has been used by us[19] and Rosner[23] to examine the consequences of mixing the ι with the η and η', and by Rosner and Tuan,[24] Schnitzer[25] and by us[19] to examine mixing the θ with f and f'.

In this work we use the simplest mass matrix of this type and the precisely known masses of the isoscalars $\eta(549)$, $\eta'(960)$, f(1273), f'(1520) to interpret the gluonium candidates. We find

(i) The $\iota(1440)$ cannot be interpreted in this framework as a gluonium candidate. With a linear mass matrix the lowest possible "gluonium" mass is 2.4 GeV; for a quadratic mass matrix it is 1.7 GeV.

(ii) The mass of $\theta(1670)$ can be accomodated with either a linear or a quadratic Fuchs mass matrix but then the observed branching ratios are in contradiction with the gluonium interpretation.

We are assuming, for the above conclusions, that the mass matrix should give the observed eigenvalues with an error of less than 1 MeV.

The mass matrix of Fuchs[20] has the form:

$$M = \begin{bmatrix} M_N + 2\alpha & \sqrt{2}\alpha & \sqrt{2}\beta \\ \sqrt{2}\alpha & M_S + \alpha & \beta \\ \sqrt{2}\beta & \beta & M_G \end{bmatrix} . \quad (1)$$

It is a generalization in the presence of gluonium of the quarkonium mass matrix of de Rujula-Georgi-Glashow[21] and Isgur,[22] which can be obtained formally from (1) in the limit of $\beta = 0$. The parameter β describes the mixing of quarkonia with gluonium; α describes the quarkonium → quarkonium annihilation amplitudes. In principle, both α and β are flavor dependent due to the different masses of quarkonia of different (hidden) flavor; in addition α and β could be related if one assumes that the transition $q_i\bar{q}_i \to q_i\bar{q}_i$ takes place through the gluonium intermediate state:

$$\alpha_i = \frac{\beta_i^2}{M_{q_i\bar{q}_i} - M_G} . \quad (2)$$

However, to retain predictive power we shall confine ourselves to the <u>simplest possibility</u>, in which α and β are unrelated <u>constants</u>. In this case the different eigenstates of the matrix (1) are orthogonal to each other. This simple limit has also been considered by Rosner[23] and Rosner and Tuan.[24]

The masses M_N and M_S are those of "nonstrange" $(u\bar{u} + d\bar{d})/\sqrt{2}$ quarkonium and "strange" $s\bar{s}$ quarkonium and are determined from isotriplet and isodoublet states of the same J^{PC}. This gives for the pseudoscalar multiplet, in a linear mass matrix:

$$0^{-+}: \quad M_N = M_\pi = 0.135 \text{ GeV}, \ M_S = 2m_K - m_\pi = 0.853 \text{ GeV} \quad ,$$

while in the tensor multiplet

$$2^{++}:\quad M_N = m_{A_2} = 1.318 \pm 0.005\ \text{GeV},\ M_S = 2m_{K^*} - m_{A_2}$$
$$= 1.55 \pm 0.01\ \text{GeV} \quad .$$

In the quadratic version,

$$0^{-+}:\quad M_N = 0.018\ \text{GeV}^2,\ M_S = 0.472\ \text{GeV}^2,$$

while

$$2^{++}:\quad M_N \simeq 1.737\ \text{GeV}^2,\ M_S \simeq 2.376\ \text{GeV}^2.$$

The parameter M_G in (1) is the "bare" gluonium mass in the absence of coupling ($\beta = 0$) to quarkonium. This is the quantity which should be compared to the results of "pure" lattice gauge theories, or bag calculations, i.e., models without any fermions. In general, when β is not zero, M_G differs from the corresponding eigenvalues $\lambda_1\lambda_2\lambda_3$ of the mass matrix, which are the physical masses in the presence of coupling to fermions. The eigenvalues λ_1, λ_2, λ_3 of M are the masses of the physical states (η, η' and a hypothetical pseudoscalar glueball G for 0^{-+}, and f, f' and a 2^{++} glueball in the tensor system). The unknowns are M_G, α, and β, and if the physical glueball mass is known , these are completely determined, as are the quark-glueball mixtures of the physical states. The three equations which determine the unknowns are:

$$\text{Tr } M = \lambda_1 + \lambda_2 + \lambda_3 = M_N + M_S + M_G + 3\alpha \tag{3}$$

$$\frac{1}{2}\left[(\text{Tr } M)^2 - (\text{Tr } M^2)\right] = \lambda_1\lambda_2 + \lambda_2\lambda_3 + \lambda_3\lambda_1$$
$$= (M_N+2\alpha)(M_S+\alpha) + M_G(M_N+M_S+3\alpha) - 2\alpha^2 - 3\beta^2, \tag{4}$$

$$\det M = (M_N+2\alpha)(M_S+\alpha)\ M_G - 2M_G\alpha^2 - (M_N+2M_S)\ \beta^2\ . \tag{5}$$

Let us first consider the pseudoscalar multiplet. We find no solution for real α, β given the set $\lambda_1 = 0.549$ GeV, $\lambda_2 = 0.960$ GeV, and $\lambda_3 = 1.44$ GeV, the quoted ι mass. As we increase λ_3, β^2 increases from its initial negative value and passes through zero at $\lambda_3 \simeq 2.4$ GeV (more precisely 2.38 GeV). This value can be obtained directly if we set $\beta = 0$ in (1) and constrain α such that one of the eigenvalues of the 2×2 submatrix is 0.549 GeV. Then, the second eigenvalue of this 2×2 matrix is 2.4 GeV, and the third eigenvalue $\lambda_3 = M_G = 0.96$ GeV. Of course, this solution is not acceptable, since the $\eta'(960)$ cannot be pure gluonium, as we know[23] from its decay $\eta' \rightarrow \rho\gamma$, which would be forbidden. This means $\beta = 0$ is not an allowed solution!

Therefore, we conclude that the $\iota(1440)$ cannot be interpreted as a glueball in this framework. The same conclusion emerges from the quadratic version of the Fuchs matrix, where β passes through zero when the third eigenvalue of the mass matrix is $(1.7\ \text{GeV})^2$. This conclusion is implicitly contained in Fuchs's work. As before, when $\beta = 0$ it is the η' which is purely gluonium and this is unacceptable, on account of, say, its radiative decays. Because of this, the actual value of λ_3 must be 2 - 2.5 GeV or even higher. Such values coincide with LTG calculations of the glueball spectrum. We also conclude that a quadratic mass matrix offers a more realistic way of treating the mixing because it leads to lower eigenvalues.

If the $\iota(1440)$ is not gluonium, what is it? Presumably, it is composed of $q\bar{q}$, $q\bar{q}q\bar{q}$ or $q\bar{q}g$. In order to determine its $q\bar{q}$ content, the radiative decays $\iota \rightarrow \rho\gamma$ or $\iota \rightarrow \rho^\circ\gamma$ are especially useful. For a predominantly $q\bar{q}$ composition one would expect a relatively large rate for at least one of these reactions. For example, if the ι has the same composition as the $\eta'(960)$, its decay rate into $\rho^\circ\gamma$ is expected to be 2 MeV.

Donoghue and Gomm[26] and Chanowitz[2(d)] have argued that the ι is not a $q\bar{q}$ radial excitation. If we accept their arguments together with the argument of this note, one might conclude that the ι has a large four quark or $q\bar{q}g$ component. It would be most interesting to have more direct evidence for this possibility.

We find that we can fit the mass value of the θ together with the f at 1.273 GeV and f' at 1.516 GeV for either a linear or quadratic mass matrix. This model fails to account for observed decay branching ratios however. In particular, branching ratios or limits on the decay of the 2^{++} system into $\eta\eta$, $\pi\pi$ and KK are known. If the glueball were unmixed or at least mixed with flavor singlets only, then ratios such as $[\Gamma(\theta\to\pi\pi)]/[(\theta\to\eta\eta)]$ would be given by symmetry limit values; the matrix elements squared would be $\theta \to \pi\pi/\eta\eta=3$ and $\theta \to \pi\pi/KK=3/4$. The salient experimental fact[2(f)] is that the ratio $[\Gamma(\theta\to\pi\pi)]/[(\theta\to\eta\eta)] <1.1$, a number which includes phase space factors. Mixing will suppress this ratio, but any such mixing must be consistent with the masses.

We use Rosner's[23,24] notation for configuration mixing, namely the eigenstates are defined by

$$|\psi_i\rangle = x_i \left|\frac{u\bar{u}+d\bar{d}}{\sqrt{2}}\right\rangle + y_i|s\bar{s}\rangle + z_i|G\rangle \ , \tag{6}$$

where i = 1 represents the f, i = 2 the f', and i = 3 the θ. For the mass matrix of Eq. (1), the values x_i, y_i, and z_i are simply determined by

$$\frac{y_i}{x_i} = \frac{1}{\sqrt{2}}\ \frac{\lambda_i-M_N}{\lambda_i-M_S} \tag{7}$$

and

$$\frac{z_i}{x_i} = \sqrt{2}\ \beta\ \frac{\left(1+\frac{1}{2}\ \frac{\lambda_i-M_N}{\lambda_i-M_S}\right)}{\lambda_i-M_G} \ , \tag{8}$$

together with the normalization condition

$$x_i^{\ 2} + y_i^{\ 2} + z_i^{\ 2} = 1. \tag{9}$$

For the quadratic version, $M_N \to M_N^2$, $M_S \to M_S^2$, $M_G \to M_G^2$, and $\lambda_i \to \lambda_i$. Thus, all the x_i, y_i, and z_i are determined from the mass fit. For the linear mass matrix, we find $M_G = 1.642 \pm 0.043$, $\alpha = -0.014 \mp 0.002$, $\beta = \pm 0.049 \pm 0.011$ and

$$x_2 = -0.169 \mp 0.001,\ y_2 = -0.946 \pm 0.006,\ z_2 = -0.277 \pm 0.024, \tag{10}$$
$$x_3 = 0.159 \pm 0.017,\ y_3 = 0.307 \mp 0.026,\ z_3 = 0.938 \pm 0.005.$$

For the quadratic mass matrix, $M_G = 1.653 \pm 0.042$, $\alpha = -0.025 \mp 0.01$, $\beta = \pm 0.174 \pm 0.040$, and

$$x_2 = -0.181 \mp 0.001,\ y_2 = 0.932 \pm 0.008,\ z_2 = -0.314 \pm 0.025, \tag{11}$$
$$x_3 = 0.192 \mp 0.027,\ \ y_3 = 0.345 \mp 0.02,\ \ z_3 = 0.919 \pm 0.001.$$

To relate these wave functions to the hadronic decay widths, we note that one new dynamical parameter r is required, which we take[24] as

$$r = \frac{\langle \pi^+\pi^- | G \rangle}{\langle \pi^+\pi^- | \frac{u\bar{u}+d\bar{d}}{\sqrt{2}} \rangle} . \tag{12}$$

This parameter and the usual quark model assignments for the K and π, and the assignment[21,22] $\eta = \frac{u\bar{u}+d\bar{d}}{2} - \frac{s\bar{s}}{\sqrt{2}}$ allow us to write ratios of partial widths, in particular

$$\frac{\Gamma(f'\to\pi\pi)}{\Gamma(f'\to K\bar{K})} = 3\left(\frac{p'_\pi}{p'_K}\right)^5 \left(\frac{x_2 + z_2 r}{x_2 + \sqrt{2}y_2 + 2z_2 r}\right)^2 \tag{13}$$

and

$$\frac{\Gamma(\theta\to\pi\pi)}{\Gamma(\theta\to\eta\eta)} = 12\left(\frac{p'_\pi}{p_\eta}\right)^5\left(\frac{x_2 + z_2 r}{x_3 + \sqrt{2}\, y_3 + 2z_3 r}\right)^2 . \tag{14}$$

p'_π is the center-of-mass momentum for the decay $f' \to \pi\pi$, etc. Our strategy is now to use Eq. (13) to determine r and then to predict the left-hand side of Eq. (14). Unfortunately, there is a sign of ambiguity in the determination of r from Eq. (13), which we resolve as in Rosner and Tuan[6] and references cited therein. In addition, we take[28] $\Gamma(f'\to\pi\pi) = 8.0 \pm 3.0 \times 10^{-4}$ GeV and $\Gamma(f'\to K\bar{K})$ to run from 0.030-0.060 GeV.

For the linear case we find for r a range of values running from -0.33 to -0.50, and for the quadratic case a range -0.32 to -0.50. These ranges fold in the errors in the vectors (x_i, y_i, z_i), in $\Gamma(f'\to\pi\pi)$ and in $\Gamma(f'\to K\bar{K})$. In turn using r and Eq. (14), we find for the ratio $R = \Gamma(\theta\to\pi\pi)/\Gamma(\theta\to\eta\eta)$, $R \geq 32$ for the linear case and $R \geq 49$ for the quadratic case. This should be compared to the experimental limit $R < 1.1$. Indeed, for the range of r allowed in our fit, it is much simpler to enhance the branching fraction for $\theta \to \pi\pi$ than to suppress it.

2^{++} glueballs in the 2-3 GeV range have been predicted[5] and possibly seen.[16] We would in general suggest that the distance of these states from the low-lying tensor nonet will make their mixing relatively small so that the glueball will behave like a flavor singlet in its decays. We have concentrated on the ratio $\pi\pi$ to $\eta\eta$ because it has given us a clear experimental test of the nature of the θ. For future glueball candidates it is useful to remember that other hadronic channels can be studied. In addition, as has been emphasized previously,[23,2(d),4(c)] radiative modes of both production and decay, and in both one-photon and two-photon couplings, are powerful probes of the glueball mixing parameters. The same considerations which lead to predictions for the hadronic decay modes of the θ also lead to values of, for example, the decay mode $\theta \to \gamma\gamma$.

IV. Open Questions

(a) Where might a 2^{++} glueball be? The 2-3 GeV range seems to be a prime region. Certainly, the LGT calculations are consistent with this possibility. The $L(2160)$ and $L(2320)$ resonances fit in nicely. For 2^{++} glueballs in the 2-3 GeV range, mixing with the low-lying f and f' would necessarily be small. so such glueballs should behave like flavor singlets in these decays.

The real situation is probably more complicated than discussed above, because of the large number of 2^{++} states that exist experimentally, and the even larger number of low-lying $q\bar{q}$ and glueball states that should exist. These states are shown below. Only a small number of these can be easily related.

2^{++} Experiment			
f(1273)	$\leftrightarrow$	$^3P_2$ (nonstrange)	
f'(1516)	$\leftrightarrow$	$^3P_2$ (strange)	
θ(1670)		$^3P_2{}^R$(nonstrange)	
(1800)		$^3P_2{}^R$(strange)	
L(2160)		$^3F_2$ (nonstrange)	
L(2320)		$^3F_2$ (strange)	
		G(2g)	L = 0, S = 2
			L = 2, S = 0
			L = 2, S = 2
		G(3g)	L = 1, S = 1
			L = 1, S = 2

The $^3P_2{}^R$ (nonstrange) has been predicted by Cohen, Isgur and Lipkin[28] to fall at 1860 MeV and possibly could be correlated with the state seen at 1800 by Cason, et al.[29] and Costa, et al.[30] Other than that no simple identifications are obvious.

(b) What is $\theta(1670)$?

1. It acts like a radial excitation of f', i.e., it acts like it is made of an $s\bar{s}$ pair. It likes to go to $\eta\eta$ and $K\bar{K}$, but not to $\pi\pi$. It is rather peculiar, however, to have what appears to be a radial excitation of the f' be lower than the radial excitation of the f, which we tentatively identify with the (1800) MeV state seen in $\pi\pi$ and $K\bar{K}$.

2. What else could it be? A 4-quark state, $s\bar{s}\,\frac{(u\bar{u}+d\bar{d})}{\sqrt{2}}$. It could fall apart to $\eta\eta$, but not to $\pi\pi$.

V. Testing Glueball Candidates

There are several tests to which glueball candidates should be subjected:

A. Radiative modes of production and decay.

B. Hadronic channel decay ratios, using SU(3).

C. 2γ couplings.

I regard the $L(2160)$ and $L(2320)$ resonances as our best candidates at present, so let's subject them to these tests.

A. $\psi \to \gamma L$? The L's are very broad. At present we can't tell whether or not they are produced in this mode.

B. We should find modes other than $\phi\phi$ where simple SU(3) tests might show whether L is a flavor singlet.

Two pseudoscalar final states: If L is a flavor singlet, then

$$\frac{\tilde{\Gamma}(L\to\pi\pi)}{\tilde{\Gamma}(L\to\eta\eta)} = \frac{3}{1} ,$$

$$\frac{\tilde{\Gamma}(L\to K\bar{K})}{\tilde{\Gamma}(L\to\eta\eta)} = \frac{4}{1} .$$

$\tilde{\Gamma}$ is the reduced width, i.e. without phase space factors. As yet, we have no evidence for these channels.

C. No evidence.

VI. Conclusions

1. $\iota(1440)$ is not a glueball, because no consistent mass matrix is formed.

2. $\theta(1670)$ is not a glueball, because branching ratios are bad.

3. $L(2160)$ and $L(2320)$ are possible glueball states, but their branching ratios must be checked.

4. We must find oddballs for the clearest indication of the existence of glueballs!

I am indebted to J. M. Cornwall, P. Fishbane, G. Karl, S. J. Lindenbaum, S. Pinsky, and A. Soni for many useful discussions and suggestions.

References

1. A number of excellent reviews of glueball properties have appeared recently. These contain extensive references to current and earlier work which may be omitted here. Some of these reviews are listed in Ref. 2.
2. (a) C.E. Carlson, Orbis Scientiae, Miami Beach, Florida, January 18-21, 1982.
   (b) Daniel Scharre, Orbis Scientiae, Miami Beach, Florida, January 18-21, 1982.
   (c) K.A. Milton, W.F. Palmer, and S.S. Pinsky, XVIIth Recontre de Moriond, March 20-26, 1982, Les Arcs, France.
   (d) M. Chanowitz, Proceedings of the 1981 SLAC Summer Institute (Stanford, California, 1981), LBL-13593, November, 1981.
   (e) Sydney Meshkov, Proceedings of the Johns Hopkins Workshop on Current Problems in Particle Theory, Florence, Italy, June 2-4, 1982.
   (f) Elliot D. Bloom, International Conference on High Energy Physics, July 26-31, 1982, Paris, France.

3. P. Fishbance, G. Karl, and S. Meshkov, UCLA preprint UCLA/82/TEP/18.
4. (a) D. Robson, Nucl. Phys. B130, 328 (1977).
   (b) J. Coyne, P. Fishbane, and S. Meshkov, Phys. Lett. 91B, 259 (1980).
   (c) J.D. Bjorken, SLAC Summer Institute on Particle Physics, SLAC-PUB-2372.
5. (a) B. Berg and A. Billoire, Phys. Lett. 113B, 65 (1982); 114B, 324 (1982).
   (b) K. Ishikawa, M. Teper, and G. Schierholz, Phys. Lett. 116B, 429 (1982).
   (c) K. Ishikawa, A. Sato, G. Schierholz, and M. Teper, Phys. Lett. 120B, 387 (1983).
6. B. Berg, A. Billoire, and C. Rebbi, BNL 30826 (preprint, December, 1981).
7. J.M. Cornwall and A. Soni, Phys. Lett. 120B, 431 (1983).
8. C. Bernard, Phys. Lett. 108B, 431 (1982).
9. G. Parisi and R. Petronzio, Phys. Lett. 94B, 51 (1980).
10. C.E. Carlson, T.H. Hansson and C. Peterson (preprint, 1982).
11. J. Donoghue, K. Johnson and B. Li, Phys. Lett. 99B, 416 (1981).
12. Catherine Newman-Holmes, SLAC-PUB-2971 (September, 1982).
13. Frank Porter, "Status of Search for Gluonia at Spear," (unpublished).
14. Melissa Franklin, SLAC Report 254, August 1982.
15. D.L. Burke, et al., Phys. Rev. Lett. 49, 632 (1982).
16. A. Etkin, et al., Phys. Rev. Lett., 40, 422 (1981); S. J. Lindenbaum, Nuovo Cimento 65A, 222 (1981); S.J. Lindenbaum, APS Meeting of DPF at the University of Maryland, College Park, Maryland, October 1982 and references cited therein; S.J. Lindenbaum, Talk at XVIII Recontre de Moriond, January 17-23, 1983, La Plagne, France.
17. T.A. Armstrong, et al., Nucl. Phys. B196, 176 (1982).
18. M. Baubillier, et al., Phys. Lett. 118B, 450 (1982).

19. Paul M. Fishbane, Gabriel Karl and Sydney Meshkov, UCLA/82/-TEP/18.

20. N. Fuchs, Phys. Rev. D14, 1912 (1976).

21. A. de Rújula, H. Georgi, and S.L. Glashow, Phys. Rev. D12, 147 (1975).

22. N. Isgur, Phys. Rev. D13, 122 (1976).

23. J.L. Rosner, preprint July 1982.

24. J.L Rosner and S.F. Tuan, preprint UH-511-477-82, 1982.

25. H.J. Schnitzer, Nucl. Phys. B207, 131 (182).

26. J.F. Donoghue and Harold Gomm, Phys. Lett. 112B, 409 (1982).

27. A.J. Pawlicki, et al., Phys. Rev. D15, 3196 (1977); M. Nikolic, University of Minnesota preprint, 1981, submitted to Phys. Rev. D.

28. I. Cohen, N. Isgur and H.J. Lipkin, Phys. Rev. Lett., 48, 1074 (1982).

29. N.M. Cason, et al., Phys. Rev. Lett. 48, 1316 (1982).

30. G. Costa, et. al. Nucl. Phys. B175, 402 (1980).

ON THE MEASUREMENT OF α_s*

L. Clavelli

High Energy Physics Division

Argonne, Illinois 60439

Abstract

We point out that a number of QCD tests, relatively free of obvious nonperturbative corrections and other theoretical problems, are now available in e^+e^- annihilation. By focusing on these tests, one can see the beginning of a confirmation of the running of the strong coupling constant predicted by the renormalization group.

QCD is a theory of strong interactions with only one free parameter, the strong fine structure constant, α_s. In principle, one measurement of α_s in any process is sufficient to determine all strong interaction processes (ignoring effects of grand unification). For instance, it has been suggested that one particularly good measurement of α_s is provided by Υ decay. The prediction of QCD, including next to leading effects[1] is

$$\frac{\Gamma(\Upsilon \to 3g)}{\Gamma(\Upsilon \to \mu^+\mu^-)} = \frac{10}{81e_b^2} \frac{\pi^2-9}{\pi} \frac{\alpha_s^3(M_\Upsilon)}{\alpha_{em}^2} \frac{[1 + (3.8 \pm .5) \frac{\alpha_s}{\pi}]}{[1 - \frac{16}{3}\frac{\alpha_s}{\pi}]} . \qquad (1)$$

*Worked performed under the auspices of the United States Department of Energy.

The three gluon decay rate is the total hadronic decay rate minus radiative and photon induced contributions, which can be deduced from off resonance measurements. The T is presumably heavy enough (9.46 GeV) to justify a perturbative calculation, and since the ratio in Eq. (1) depends on the third power of α_s, it provides a very sensitive measure of the strong coupling. The conclusion of Mackenzie and Lepage is

$$\alpha_s(M_T)_{\overline{MS}} = 0.141 \begin{matrix} +.009 \\ -.008 \end{matrix} \quad , \tag{2}$$

or perhaps better

$$\alpha_s(0.48\ M_T) = 0.158 \begin{matrix} +0.012 \\ -0.010 \end{matrix} \quad . \tag{3}$$

The scale 0.48 M_T is that at which the next to leading order corrections are absorbed into the leading α_s^3 term. Given a measure-emt of α_s at one energy, such as the above, the renormalization group predicts its value at every other energy according to a linear differential equation

$$x = \frac{\alpha_s(q^2)}{4\pi} \quad , \tag{4}$$

$$t = \ell n\ q^2/\Lambda^2 \quad , \tag{5}$$

$$\frac{dx}{dt} = =ax^2(1 + b_1 x + b_2 x^2 + \ldots) \quad , \tag{6}$$

where, for n_f quark flavors,

$$a = 11 - \frac{2n_f}{3} \quad , \tag{7}$$

$$b_1 a = 102 - \frac{38}{3} n_f \quad , \tag{8}$$

and (in the MS scheme)[2]

$$b_2 a = \frac{2857}{2} - \frac{5033}{18} n_f + \frac{325}{54} n_f^2 . \qquad (9)$$

At present energies ($n_f < 6$) the solution of Eq. (6) is

$$\ell n\, q^2/\Lambda^2 = \frac{1}{ax} [1 + \frac{b_1}{2} x \;\; \ell n[\frac{x^2}{1 + b_1 x + b_2 x^2}]$$
$$- \frac{[b_1^2 - 2b_2]}{\sqrt{4b_2 - b_1^2}} x \tan^{-1} [\frac{x\sqrt{4b_2 - b_1^2}}{2 + b_1 x}] + \ldots]. \qquad (10)$$

(It is interesting to note that for $n_f > 6$ the arctan in Eq. (10) is replaced by an arctanh and the β function develops an apparent second zero.[3]

A major goal of strong interaction physics should be the experimental confirmation of the variation of α_s with q^2 according to Eq. (10). Until this property of asymptotic freedom is clearly observed it cannot be said that QCD is experimentally established.

The decrease of α_s with increasing energy is thought to be related to phenomena[4] such as the slow angular narrowing of jets observed in e^+e^- annihilation and to the increasing planarity of events with increasing energy,

$$\frac{\Sigma|P_T|}{\Sigma|P_{\parallel}|} \xrightarrow[s \to \infty]{} 0 \quad , \qquad (11)$$

$$\frac{\Sigma|P_T|_{out}}{\Sigma|P_T|_{in}} \xrightarrow[s \to \infty]{} 0 \quad . \qquad (12)$$

Unfortunately, most of the presently availabe data are so plagued by nonperturbative corrections and/or systematic errors that the decrease of α_s with energy has not been clearly demonstrated. This

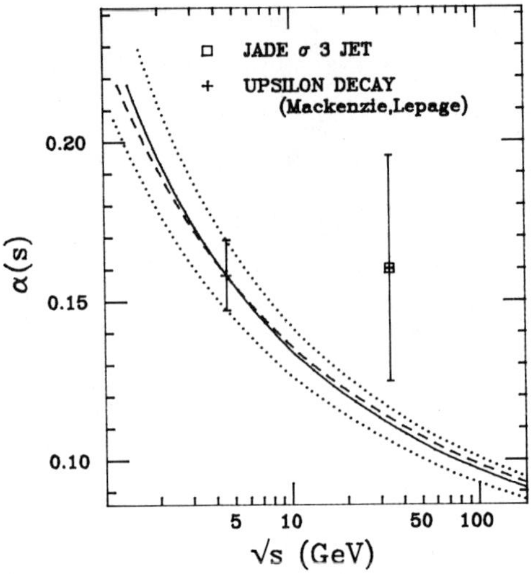

Fig. 1. Jade measurement of α_s compared to the renormalization group extrapolation from the apparent value in upsilon decay.

is exemplified in Fig. 1 where the Υ decay result of Eq.(3) is extrapolated to higher (and lower) energies via Eq.(10) and compared to the most recent high energy measurement:[5] $\alpha_s(33.8) = .16 \pm .015(\text{stat}) \pm .03(\text{syst})$. The slope of the curves changes discontinuously at flavor thresholds according to Eqs.(6-9). The open top threshold is taken to be 50 GeV. The dashed line shows the lowest order ($b_1 = b_2 = 0$) renormalization group extrapolation. The second order extrapolation ($b_2 = 0$ in Eq.(6) is indistinguishable within the resolution of Fig. 1 from the cubic order solution shown in the solid and dotted lines. Because of the large errors, the Jade result is consistent with the extrapolation from the Υ but certainly cannot be used to support the running of the coupling constant. Indeed, if one extrapolates $\alpha_s(33.8) = 0.16$ down to the Υ mass one predicts $\alpha_s(M_\Upsilon) = .223$ which, inserted into Eq. (1), overestimates the Υ 3 gluon to μ pair ratio by more than a factor of 5. Clearly, if $\alpha_s(30 \text{ GeV}) \approx .16$, as many results seem to indicate,[6] perturbative QCD has little to do with Υ decays.

In most of these tests, however, nonperturbative effects are very important. One must remember that no model for nonperturbative corrections is more than a model. If a good fit to an experimental distribution with 50% model dependent corrections yields a particular value for α_s, no one should doubt that there exists another fragmentation model yielding an equally good fit with a widely different result for α_s. Precisely, this has been demonstrated, in fact, by the Cello group[7] fitting the energy-energy correlation using two different fragmentation models.

Fortunately, it is, in principle, possible to determine quantitatively the amount of nonperturbative contamination by observing the energy dependence of jet measures. Fragmentation effects are higher twist terms in strong interaction theory and must therefore fall as a power of the center of mass energy, $\sqrt{s}$, in e^+e^- annihilation, whereas perturbative QCD contributions fall as a power of ℓn s. This crucial study of energy dependences has been performed by the PLUTO group, some of whose results will be discussed later.

In addition to checking energy dependence, it is important to search for jet observables whose distribution in the perturbative tail follows the shape of the pure (parton level) QCD. The existence of such variables was the original promise of the idea of infrared insensitive observables. This ideas has proven, however, to be an elusive dream as is well illustrated by the case of thrust, T, which is the total jet 3 momentum resulting from the association of final state particles to two jets in the way that maximizes this three momentum. In Fig.2 is shown the data of Mark J compared to the lowest order QCD prediction (upper solid line) for $\alpha_s = 0.17$. The data was originally well fit by this curve together with an obviously large contribution from a nonperturbative hadronization model. In 1980 the second order QCD matrix elements were calculated[8] which enabled one to draw the α_s^2 contribution[9] to the thrust distribution shown, for $\alpha_s = 0.17$, in the upper dashed curve of Fig.2. The total theoretical thrust distribution to this order would be the sum of these two curves except that a prudent theorist

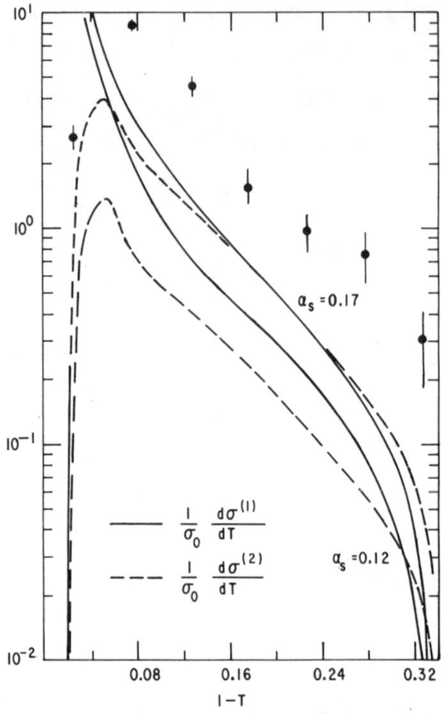

Fig. 2. Experimental thrust spectrum compared to first order QCD contribution for α_s = 0.17 (upper solid curve), 2nd order QCD contribution for α_s = 0.17 (upper dashed curve), first order QCD for α_s = 0.12 (lower solid curve), and 2nd order contribution for α_s = 0.12 (lower dashed curve).

hesitates to add terms in a perturbation series when the higher order corrections are as big or larger than the leading order.

One approach to the large second order corrections is to revise downward one's value of α_s whereas the second order is quadratic in α_s, so the apparent convergence is improved. The sum of the 1st and the 2nd order is then equal to the old 1st order curve allowing, presumably, a good fit to the data after including the large fragmentation corrections, and the value of α_s is then equal to that predicted in Fig. 1 from the renormalization group extrapolation from T decay.

If one insists on keeping α_s (30 GeV) ≈ 0.17, it is clear that pertubative QCD has little or nothing to do with the thrust distribution and that the nonperturbative corrections are not as well understood as one thought two years ago, both of which may well be true. In addition, whether α_s is small or large, it is clear that nonperturbative corrections are very important since the shape of the pure QCD is not followed by the data in the small thrust (perturbative) tail. This is true of most jet distributions that have been measured and it has been suggested[10,11] that a common cause is a receding endpoint effect whereby final states with higher multiplicity (of pions or partons) have a larger kinematic range for the observable in question. This causes the higher order terms in the perturbation expansion to exceed eventually the lower order terms as seen in Fig. 2. This phenomenon is observed in thrust, spherocity, acoplanarity and most of the other jet measures.

The result is, first of all, a slow convergence of the perturbation expansion in the so called "perturbative tail", and secondly large nonperturbative corrections in the tail region since it is kinematically "easier" for a multihadron state to populate the tail region than for a few parton state.

One way out of this problem, which has been favored by many is to define a large part of the experimental events in the tail region and of the order α_s^2 QCD corrections to be "4 jet events" and to exclude them from the analysis. Usually this is done by a cluster algorithm which decides whether a given event is two jets, three jets or four jets, etc.

Having defined a sample of pure 3 jet events, one can construct variables which do not have the receding endpoint effect discussed above. For example, one can consider the scaled jet energies $x_i = \frac{2E_i}{\sqrt{s}}$. Clearly, independent of the multiplicity within each jet, $x_{max} > 2/3$. To first order in α_s, x_{max} = thrust; it is sometimes called "cluster thrust". In practice, one does not measure the actual jet energies, but the energies the jets would have if they were massless but separated by the observed angles.

$$x_i = \frac{2 \sin\theta_{k\ell}}{(\sin\theta_{12} + \sin\theta_{23} + \sin\theta_{31})} , \quad i,k,\ell \text{ cyclic} . \tag{13}$$

The α_s(33.8 GeV) given in Ref.5 and shown in Fig. 1 was determined by measuring the distribution in this x_{max} and in an analogously defined x_1.

The parameters in the cluster algorithm can be chosen to make the x_{max} distribution of the three jet sample follow the shape of the lowest order QCD, allowing a fit to α_s seemingly free of non-perturbative uncertainties. The cluster algorithms used in Ref. 5 throw out only about 4% of the data as 4 jet events or higher, but this 4% is probably a significant fraction of the perturbative tail. That is, a relatively small variation in the parameters of the cluster algorithm can cause a significant change in the number of events in the x_{max} tail. This may be reflected in the large systematic errors (twice the statistical error) in the JADE measurement.

There are also theoretical problems discussed in Ref. 5, in including second order QCD corrections. One might worry that the cluster algorithm couples more strongly to the pion level than to the parton level. The second order QCD study of the x_{max} variable has been mostly pursued by G. Kramer and G. Schierholz[12]. They use two jet resolution methods. The first is an energy-angle type definition and the second is a definition based on a minimum resolvable jet invariant mass. Both methods correspond only loosely to the cluster algorithms used in the data against which the theory is compared. The two methods yield different values of α_s, and in each case the resulting value of α_s is a function of the resolution parameters.[12] Although by a judicious adjusting of cluster parameters and jet resolution parameters, the data can be made to follow the shape of the bare QCD, one is, perhaps, not left with a great feeling of confidence that α_s can be accurately measured by this method.

In the search for a good QCD variable one might, therefore, be wise to consider two criteria:

C1) Seek a variable whose kinematic endpoint is multiplicity independent (to reduce higher order QCD and nonperturbative effects).

C2) Avoid variables whose definition requires a decision as to whether a given event is 2 jets, 3 jets, 4 jets, etc. (to reduce systematic errors). A theory with one free coupling should not be made a function of extra parameters.

Following this path, one encounters next the energy-energy correlation proposed by the Seattle group[13]

$$f(\theta) = \frac{1}{\sigma_0} \frac{d\Sigma}{d\cos\theta} = \sum_{N=2}^{\infty} \int \frac{d\sigma(e^+e^- \to N \text{ particles})}{\sigma(e^+e^- \to \mu^+\mu^-)} \sum_{i,j=1}^{N} \frac{E_iE_j}{s}$$

$$\delta(\cos\theta_{ij} - \cos\theta) \quad . \tag{14}$$

The integral in (14) is over the full N particle phase space. It is clear that $f(\theta)$ is infrared insensitive because of the bilinear energy weighting and trivially satisfies the two criteria C1 and C2 proposed above.

Experimentally, it is indeed found that $f(\theta)$ exhibits markedly less nonperturbative smearing than does the thrust distribution. This can be further reduced by the following observation. The dominant nonperturbative effect comes from the fragmentation into hadrons of the $q\bar{q}$ parton final state. This fragmentation, however, is symmetric in $\theta \to \pi-\theta$. We may, therefore, construct the "Energy Energy Asymmetry"

$$A(\theta) \equiv f(\theta) - f(\pi-\theta) \quad , \tag{15}$$

which should have only much smaller fragmentation effects in the large θ perturbative tail. We would, therefore, expect that, in the perturbative tail (θ near 90°), $A(\theta)$ would follow the bare

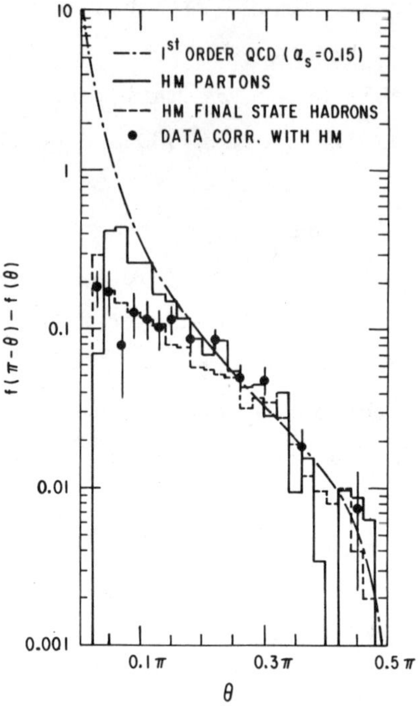

Fig. 3. CELLO results for the energy energy asymmetry compared to first order QCD (from Ref. 7).

QCD, allowing a good measurement of α_s.

The CELLO data[7] shown in Fig. 3, compared to order α_s predictions, indicates that this is indeed true. A fit to 1st plus 2nd order QCD results[14] in a lower value of α_s but does not change the impressive conclusion that bare QCD predicts the shape (and the magnitude if one takes Λ from a fit to Eq. 1) of the energy asymmetry. Within errors, no evidence for nonperturbative corrections is found (for $\theta > \frac{\pi}{5}$) and a reasonable fragmentation model (the Hoyer model) indeed gives no significant correction in this perturbative tail region. The fit of Ref. 14 is shown in Fig. 4. The resulting values of α_s are plotted in Fig. 5 and found to be in good agreement with the renormalization group extrapolation from the Υ decay result.

Figures 4 and 5 could be a significant success of QCD since no model dependent phenomological corrections are indicated or required. One can of course find a fragmentation model (or construct many others) which provides a fit to the data with other values of α_s

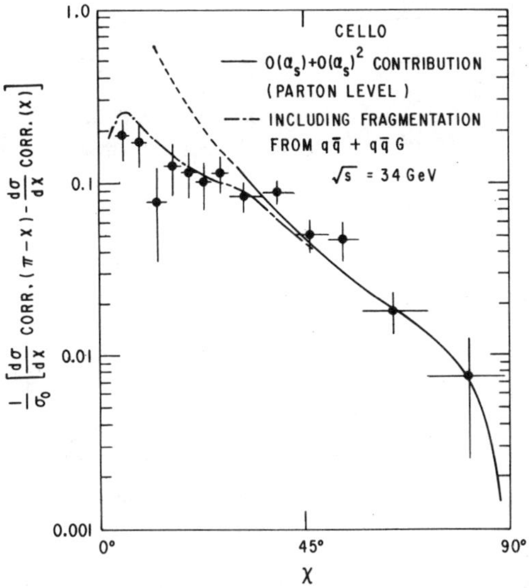

Fig. 4. CELLO results for the energy energy asymmetry fit to first plus second order QCD (from Ref. 14).

(the Lund model fit[7] indeed gives much larger value of α_s while not affecting the shape of the asymmetry). Perhaps one should be wary of fragmentation models which contain so much QCD related physics (such as string model ideas) because of possibilities of double counting when these are superimposed on the QCD perturbation theory.

The errors on α_s(30 GeV) in Fig. 5 are large partly due to the obvious reduction in statistics when one forms the asymmetry of Eq. 15. There is also, however, the following relatively minor systematic problem. In order to maximize experimental acceptance, it is customary to select jets at wide angles to the beam axis. The measurement of the perturbative tail (small cos θ) of the EE correlation requires the accurate detection of particles near 90° from the jet axis. This causes an acceptance problem since efficiencies are small near the beam directions. If the jets were azimuthally symmetric, a correction could be more easily made but QCD jets are predicted to become increasingly planar at high energies. The problem is not severe but leaves open the possibility that A(θ) is still not the ideal jet variable.

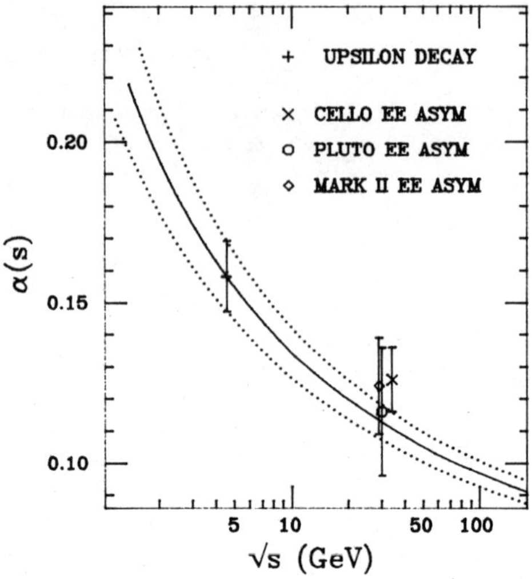

Fig. 5. High energy values for σ_s, extracted in Ref. 14 from three experiments fit to pure perturbative QCD, compared to the renormalization group extrapolation from upsilon decay.

We turn finally to various jet invariant mass measurements suggested a number of years ago[15]. Traditionally, the first question a physicist asks on observing a striking cluster of particles (jet) is its mass spectrum, but published data for e^+e^- jets is only now becoming available.

In Ref. 15 the following jet mass measurements were proposed:

1) Divide each event into two jets in the way that minimizes the sum of the two squared invariant masses.

2) Measure $\frac{1}{\sigma_0}\frac{d}{dM_H^2/s}$ and $\frac{1}{\sigma_0}\frac{d\sigma}{dM_L^2/s}$ where by definition in each event $M_H^2 \geqslant M_L^2$, H and L standing for heavy and light, respectively.

Away from the infrared region, ($M_i^2 \cong 0$), the QCD predictions are

$$\begin{aligned} \frac{1}{\sigma_0}\frac{d\sigma}{dM_H^2/s} &= \frac{\alpha_s}{\pi} f_1(M_H^2/s) + \left(\frac{\alpha_s}{\pi}\right)^2 f_2(M_H^2/s) + \ldots, \\ \frac{1}{\sigma_0}\frac{d\sigma}{dM_L^2/s} &= \left(\frac{\alpha_s}{\pi}\right)^2 g_2(M_L^2/s) + \left(\frac{\alpha_s}{\pi}\right)^3 g_3(M_L^2/s) + \ldots \quad . \end{aligned} \tag{16}$$

The first moments are predicted[10,11,15] to satisfy

$$\langle \frac{M_H^2}{s} \rangle = 1.05 \frac{\alpha_s}{\pi} + 6.9(\frac{\alpha_s}{\pi})^2 + \ldots \quad , \tag{17}$$

$$\langle \frac{M_L^2}{s} \rangle = 4.0(\frac{\alpha_s}{\pi})^2 + \ldots \quad . \tag{18}$$

In fitting data it is probably better to use the Pade approximation to (17)

$$\langle \frac{M_H^2}{s} \rangle = \frac{1.05\ \alpha_s/\pi}{1 - 6.6 \frac{\alpha_s}{\pi}} \quad . \tag{19}$$

The first nontrivial contribution to the light jet mass spectrum occurs at order α_s^2 in QCD since the final state must have at least four partons to give a nontrivial light jet mass. This leads to the prediction of a distinctive hierarchy in jet masses as $s \to \infty$,

$$\frac{\Lambda^2}{s} \ll \langle M_L^2/s \rangle \ll \langle M_H^2/s \rangle \ll 1 \quad . \tag{20}$$

It was speculated in Ref. 11 and proven in Ref. 10 that to any order in QCD (i.e. for any number of final state relativistic particles) the heavy jet mass spectrum satifies the kinematic bound

$$\frac{M_H^2}{s} \leq \frac{1}{3} \, . \tag{21}$$

The same is of course true for the distribution in the mass difference $\frac{M_H^2}{s} - \frac{M_L^2}{s}$. These two distributions satisfy therefore the criteria C1 and C2.

The 1st and 2nd order QCD predictions[10] for $\frac{M_H^2}{s}$ using the 2nd order matrix elements of Ref.8 are shown in Fig. 6. Unlike the case of thrust and such variables, the 2nd order piece lies everywhere

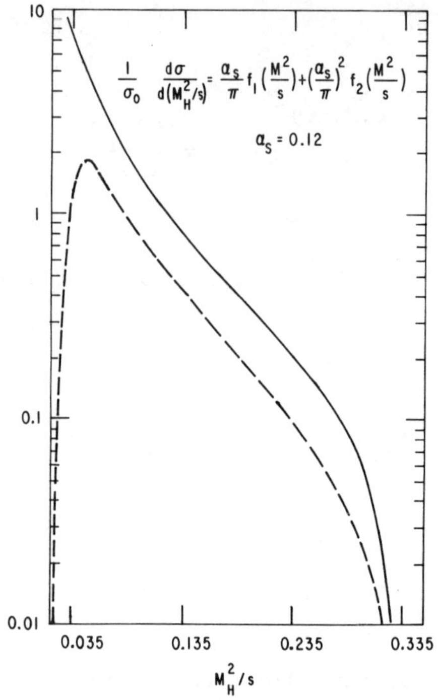

Fig. 6. First order, (solid curve) and second order (dashed curve) contributions to the mass spectrum of the heavier jet in perturbative QCD for α_s = 0.12 (from Ref. 10). The full cross section to this order is the sum of the two curves shown.

below the first order. It is interesting to compare the ratio of the first two contributions as a function of the observable for mass and 1-T. That is we write

$$\frac{1}{\sigma_0}\frac{d\sigma}{dv} = \frac{\alpha_s}{\pi} f_1(v)\left[1 + \frac{\alpha_s}{\pi} R(v)\right]. \tag{22}$$

R(v), where v is either $\frac{M_H^2}{s}$ or 1-T, is shown in Fig. 7. For thrust there is a large region over which R(v) is infinite. If we include in R(v) the cubic and higher order QCD corrections, this region of infinity extends to T = 1/2. The striking fact that the second order contribution to the mass spectrum is, to the available accuracy,

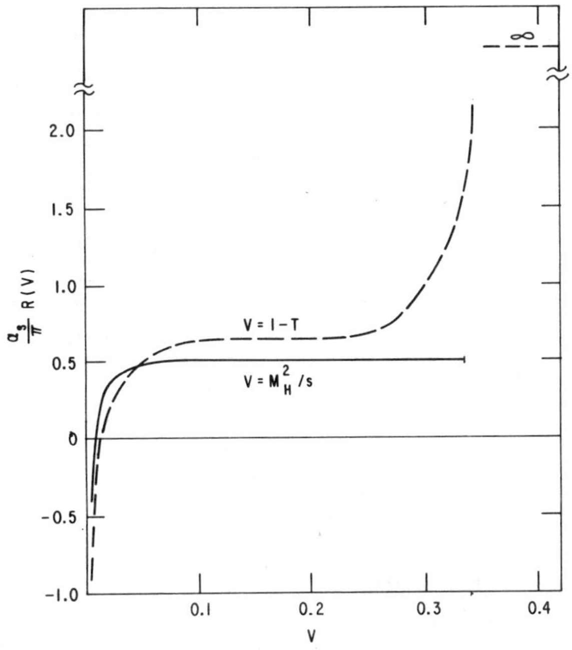

Fig. 7. The ratio, R(v), of the second order QCD contribution to the leading order contribution for v = 1 - T (dashed curve) and $v = M_H^2/s$ (solid curve). The receding endpoint effect in the thrust distribution manifests itself as a singularity in R(1-T).

proportional to the first order suggests that the higher orders may be largely absorbable into the leading order by a redefinition of the energy scale at which the coupling constant is probed.

It was shown in Ref. 10 that, if minimizing the sum of the squared jet masses defines jets J_1 and J_2 with four momenta P_1 and P_2, and, if k is the four momentum of any particle in jet 2, then

$$k\cdot(P_1 - P_2 + k) \geq 0 \quad . \tag{23}$$

This implies that the angle θ between $\vec{k}$ and $\vec{P}_2$ is restricted by

$$z_1 - z_2 + \beta \cos\theta[(1 + z_1 - z_2)^2 - 4z_1]^{1/2} + \frac{m_k}{\sqrt{s}}(1 - \beta^2)^{1/2} \geq 0. \tag{24}$$

where $z_i = P_i^2/s$ and $\beta = |\vec{k}|/|k_0|$. Equation 24 can be used to construct a fast algorithm to determine the jet definition. It also implies that for jet masses in the far perturbative tail the particles of the heavy jet are contained in a cone of half angle near 60°. This suggests that the jet mass spectrum may have smaller acceptance corrections than the energy energy asymmetry.

The first experimental study of jet mass spectra was made by E. Elsen of the JADE group and appears in his University of Hamburg dissertation.[16] Data from his thesis is shown in Fig. 8, fit to 1st plus 2nd order QCD. Over a wide range of the perturbative tail ($\frac{M_H^2}{s} > 0.15$) the data follow the shape of the bare QCD and yield[16]

$$\alpha_s(30\text{ GeV}) = 0.11 \pm 0.01. \qquad (25)$$

No evidence for significant nonperturbative effects is seen in this region and these are shown[16] to be small in a particular (Hoyer) fragmentation model. This value of α_s is compared to the renormalization group extrapolation from Υ decay in Fig. 9. The average mass squared is found to be

$$\left\langle\frac{M_H^2}{s}\right\rangle_{exp} = 0.06 \pm 0.01\ , \qquad (26)$$

which inserted into Eq. (19) yields $\alpha_s(30\text{ GeV}) = 0.130 \pm .015$. The fit to the perturbative tail of the mass spectrum yields (Eq.25) an $\alpha_s(30\text{ GeV})$ in excellent agreement with the renormalization group extrapolation from the Υ region and the EE asymmetry results shown in Fig. 5.

One important caveat is that Elsen defined the jets by a plane perpendicular to the thrust axis instead of by the combinatoric definition proposed in Ref. 15. Such a definition may not obey criterion C1 and has presumably slightly different 2nd order corrections.

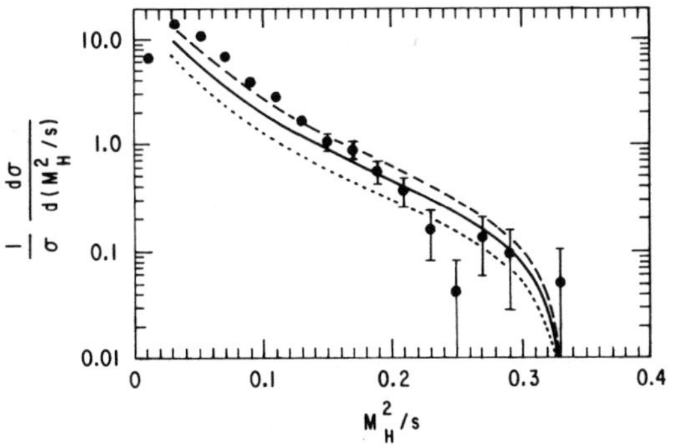

Fig. 8. The heavy jet mass spectrum from Ref. 16 compared to first plus second order pure perturbative QCD for $\alpha_s = 0.08$ (dotted curve), $\alpha_s = 0.11$ (solid curve), and $\alpha_s = 0.14$ (dashed curve). Pure QCD provides a good fit in the perturbative region of large jet mass.

There now exist also some published results on mass spectra using the combinatoric definition from the PLUTO group.[17] Their statistics are poorer than those of Ref. 16 and their mass spectra are therefore limited to $\frac{M_H^2}{s} \lesssim 0.22$. They find

$$\left\langle \frac{M_H^2}{s} \right\rangle_{exp} = 0.074 \pm .005 \quad , \tag{27}$$

for s ≅ 30.8 GeV. The hadronization studies of Elsen indicate that nonperturbative effects are small above $M^2/s = 0.2$. It would be extremely useful to have a high statistics spectrum covering the high mass tail.

Most importantly, the PLUTO group has, however, data over a range of energies from 7.7 to 31.6 GeV. The average heavy jet M^2/s shows evidence for a higher twist power law component in the energy dependence although this is appreciably smaller than that observed in the average 1-T and average jet opening angle obtained from the energy correlation measurements. This residual hadronization effect in the average $\frac{M_H^2}{s}$ is presumably due to the contribution from the

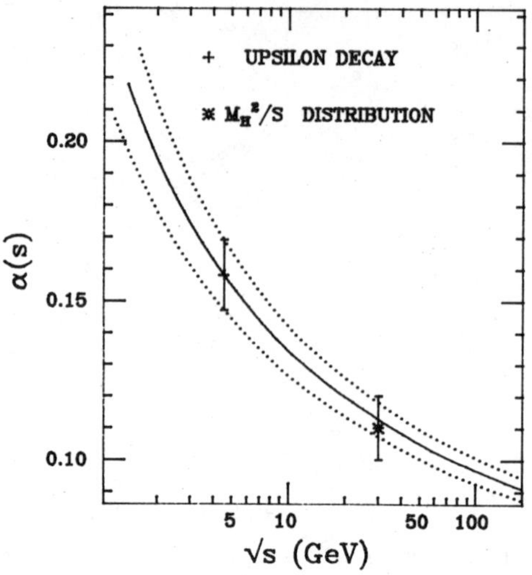

Fig. 9. High energy value for α_s, extracted in Ref. 16 from the QCD fit to the tail of the heavy jet mass spectrum, compared to the renormalization group extrapolation from the upsilon region.

the infrared region of small mass.

As in the energy asymmetry it is possible to eliminate all or most of the fragmentation effects by working with the differences of the jet squared masses, $\frac{M_H^2 - M_L^2}{s}$. The dominant fragmentation effects, coming from the parton level $q\bar{q}$ state, cancel in the difference since this state tends to hadronize symmetrically.

The QCD prediction from 17 and 18 is

$$\left\langle \frac{M_H^2}{s} - \frac{M_L^2}{s} \right\rangle = 1.05 \frac{\alpha_s}{\pi} + 2.9\left(\frac{\alpha_s}{\pi}\right)^2 \quad . \tag{28}$$

One sees that, not only are the nonperturbative effects expected to largely cancel in this difference but the higher order perturbative contributions also tend to cancel. The PLUTO data on this average as a function of energy show no evidence for a power law component.

A fit to the pure QCD results of Eq. (28) yields the α_s values

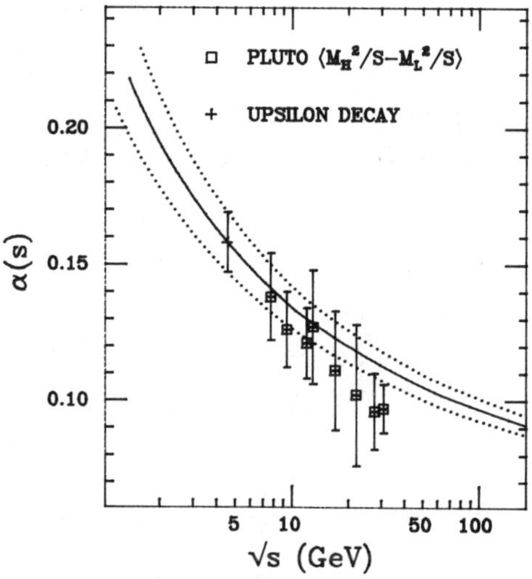

Fig. 10. Values of α_s extracted from PLUTO's average jet mass difference at eight energies compared with the renormalization group prediction from upsilon decay.

shown in Fig. 10. Agreement with the renormalization group predictions from Υ is excellent.

In conclusion we note that the energy energy asymmetry and various jet mass measurements show evidence of freedom from non-perturbative corrections. The resulting values of α_s agree with each other and with renormalization group expectations based on Υ decay. They also agree with theoretical predictions from QCD sum rules and lattice predictions and with some, but not all, of the deep inelastic scattering results. Some understanding, based on criteria C1 and C2, may be had as to why these measurements may be more reliable than others. It is clear that these relatively low values of α_s would sharply reduce the unification mass in minimal SU(5) and put this theory in gross contradiction with present measurements of proton stability.

As a word of caution, we note that analyses of systematic errors have not yet been published in the jet mass and EE asymmetry measurements (errors quoted are statistical only). We expect that systematic errors should not be a serious, ultimate obstacle to an accurate measurement of α_s through the jet masses if the combinatoric definition of the jets is adopted. We are, however, also aware that many experimentalists feel, based on many Monte Carlo studies of fragmentation, that the α_s(34 GeV) plotted in figure 1 represents our present best knowledge.

Hopefully, further study of the mass and energy correlation variables with the higher statistics now available will show whether we are really seeing in graphs such as Fig. 10 the first confirmation of the running strong coupling constant.

REFERENCES

1. P.B. Mackenzie and G.P. Lepage, Phys. Rev. Lett. 47, 1244 (1981).
2. O.V. Tarasov, A.A. Valadimirov, and A. Yu Zharkov, Phys. Lett. 93B, 429 (1980).
3. The amusing exercise of following the β function solution and this second zero as a function of n_f was carried out in collaboration with D. Wyler.
4. For a review of experimental results see K.H. Mess and B.H. Wiik DESY preprint 82-001 (1982).
5. Jade Collaboration, W. Bartel et al., Phys. Lett. 119B, 239 (1982).
6. See table 7.4 of Ref. 4.
7. Cello Collaboration, H.J. Behrend et al., Saclay preprint D Ph PE 82-08.
8. R.K. Ellis, D.A. Ross, A.E. Terrano, Phys. Lett. 45, 1226 (1980) and Nucl. Phys. B178, 421 (1981).
9. Z. Kunszt, Phys. Lett. 99B, 429 (1981) and Nucl. Phys. B178 421 (1981); L. Clavelli, D. Wyler, Phys. Lett. 103B, 383 (1981); R.K. Ellis and D.A. Ross, 106B, 88 (1981); A. Ali, DESY report 81-051. See also the calculation of J.A.M. Vermaseren, K.F.J. Gaemers and S.J. Oldham, Nucl. Phys. B187, 301 (1981).

10. L. Clavelli and D. Wyler, Phys. Lett. 103B, 383 (1981).
11. T. Chandramohan and L. Clavelli, Phys. Lett. 94B, 409 (1980).
12. G. Kramer, DESY Report 82-029; G. Schierholz, DESY report 81-042.
13. C.L. Basham, L.S. Brown, S.D. Ellis, S.T. Love, Phys. Rev. D17, 2298 (1978).
14. A. Ali and F. Barreiro DESY preprint 82-033; D. Richards, W.J. Stirling, S.D. Ellis DAMPT preprint 82/18 (1982).
15. L. Clavelli and H.-P Nilles, Phys. Rev. D21, 1242 (1980). For a review see K. Hagiwara, Madison (Wisc.) preprint MAD/PH/56 (1982).
16. E. Elsen, Hamburg dissertation, DESY F22 81/02. See also the report of A. Marshall, (Proceedings of the International Conference on High Energy Physics, Lisbon, Portugal, 1981) for a first order QCD fit to the JADE mass spectrum.
17. Pluto collaboration, Ch. Berger et al., Z. Phys. C12, 297 (1982).

B-L VIOLATING SUPERSYMMETRIC COUPLINGS

P. Ramond

Department of Physics, University of Florida

Gainesville, Florida 32611

We consider two problems: One is the possible effect of the breaking of Peccei-Quinn[1] symmetry on the inflationary universe scenario; the other is the remark that even the minimal supersymmetric SU_5 theory contains B-L violating couplings which give rise to neutrino masses and family-diagonal proton decay. However, the strength of these couplings is limited by the gauge hierarchy.

This talk will consist of two pieces: One concerns itself with the influence of Peccei-Quinn[1] symmetry breaking on the inflationary universe; the other with the role of B-L violating couplings that occur even in the minimal supersymmetry GUT theory such as SU_5.

There are at present few problems with the standard $SU_3 \times SU_2 \times U_1$ model of the low energy interactions in the sense that it has always (so far) been vindicated when confronted with experiment. So it is important to focus on whatever source of conflict there might be. The only source of disharmony with experiment comes from the fact that QCD, through the contribution of instantons, violates CP invariance. Experimentally, the absence (to the measured accuracy) of

*Supported in part by the U.S. Department of Energy under contract No. DE-AS05-81-ER40008.

an electric dipole moment for the neutron indicates that if there is any CP-violation in the strong interaction, it must be very small. Many reasons for the smallness of, or the absence of, CP-violation in QCD have been suggested. One school thinks that the smallness of the effect can be understood as a consequence of a more general theory yet to be found.[2] Their reason is that the renormalization of the effect occurs very deep in perturbation theory and that one should not worry about it until one has a theory that describes physics at very short distances. The other school, led by Peccei and Quinn[1], noted that the presence of a quark-chiral global phase symmetry in the theory would solve the problem. However, this quark-chiral symmetry is broken spontaneously by Higgs and explicitly by instanton effects. It results in a pseudo-Nambu-Goldstone boson called the axion[3] whose properties are tied to the scale at which the PQ symmetry is spontaneously broken. Through the detective work of many researchers, this mass scale has been bracketed to be between 10^{12} and 10^{8} GeV. These bounds come from astrophysical considerations: the upper bound is obtained by requiring that the axion not dominate the mass of the universe,[4] the lower one comes from the traditional picture of stellar evolution applied to red giants.[5] A much lower scale for the PQ breaking has been ruled out by looking for the axion in accelerators and not finding it.[6] Thus, if the axion exists at all, the PQ symmetry from which it stems must be spontaneously broken at ~ 10 GeV. What use is this information? It was pointed out recently that such an axion could play the dominant role in galaxy information.[7] Here, we would like to explore another possible explanation - its role vis-a-vis the first order GUT-Guth[8] transition of Grand Unified theories.

Presumably, the PQ symmetry breaking corresponds to a second order phase transition since it is a global symmetry. The GUT transition is thought to occur at a much larger temperature, but if it is first order, as in the inflationary universe picture, it will reach by supercooling temperatures of the order of T_{PQ} ($\sim 10^{10}$ GeV) before being completed. In the following we want to point out what role the PQ phase transition can play in helping or hindering the standard GUT transition.

There is already one known example[9] where a second order phase transition can dramatically affect a first order phase transition: the breaking of the electroweak symmetry is achieved by two different mechanisms, at ~ 10^2 GeV via the Higgs mechanism, and at ~ 1 GeV QCD-induced chiral symmetry breaking. As the first order phase transition with $T_c \sim 10^2$ GeV proceeds, it supercools down to 1 GeV where the QCD-induced chiral symmetry breaking induces a term linear in the Higgs field which can destroy the hump in the free energy which causes supercooling to occur. As a result, it can terminate abruptly the supercooling by effectively changing the order of the phase transition. In this case the linear term turns out to be capable of destroying the hump when the Higgs mass in the vicinity of the Coleman-E. Weinberg value. Can we have a similar effect between the GUT and PQ breaking?

To be specific, consider a single SU_5 potential consisting of a Higgs field H transforming as a 24 and a complex scalar field S carrying a nonzero PQ number. [This field could also carry supersymmetry in more sophisticated models.] The potential is (at zero temperature)

$$V = V_{24}(H) + V_1(S) + \frac{\lambda}{2} S^\dagger S \, H_{24} H_{24},$$

where V_{24} is the standard SU_5 potential and $V_1(S)$ is a potential including only the field S.

Supercooling occurs because the curvature of the free energy at the unstable point is positive, i.e.

$$\left.\frac{d^2V}{dH^2}\right|_{H=0} = \left.\frac{d^2V_{24}}{dH^2}\right|_{H=0} + \lambda S^+S > 0. \qquad (T\neq 0)$$

Assume that V_{24} itself has positive curvature at H=0. When S acquires its vacuum value at 10^{10} GeV, this can be changed if $\lambda > 0$. Thus, we see that this extra term is capable of terminating the GUT first order phase transition. In particular, if we apply this to the original Guth scenario, we see that it obviates the difficulties since the transition is abruptly terminated at 10^{10} GeV. However, this might not be completely satisfactory since not enough entropy may have been generated by such a small amount of supercooling. In this case the tremendous entropy generation would have to come from an earlier time, leaving us with no understanding for the scarcity of monopoles.

Another possible use this term might have is in the improved inflationary[10] mechanism. There one needs a region where V does not show much curvature. If V_{24} itself is set up to have such a behavior, it can be offset by the presence of the extra term which gives additional curvature to the H-field in the form $\lambda < S^+S >_0 H_{24}H_{24}$. Alternatively, it could be that the flatness required by the new inflationary scenario comes about only because of this term which cancels any curvature present in V_{24}. Finally, it affects the rate of inflation through the constraint that there be no cosmological term <u>after</u> both H_{24} and S assume their vacuum values. We hope to present a detailed account of these possibilities elsewhere.

We now turn to the second part of this talk, where we want to examine what new opportunities supersymmetric GUTs offer in our understanding of the fermion masses and mixing angles. Let us first remind the reader of the basic tools used in building such theories.[11] One starts with a chiral superfield Φ containing a left-handed Weyl fermion f_L and a complex spinless field h. It is sometimes useful to include also a dummy spinless field F. Supersymmetry invariants are constructed out of polynomials in Φ. If we schematically set

$$\Phi = (f_L, h) \ , \quad \Phi' = (f'_L, h') \ , \quad \Phi'' = (f''_L, h'') \text{ etc } \ldots,$$

then the most general N=1 supersymmetric-invariants (up to surface terms) that can appear in a Langrangian are

$$m\int d^2\theta\Phi\Phi' \to m\ f_L^T\sigma_2 f_L + m^2(h^*h + h^{*\prime}h') \quad ,$$

and/or

$$\lambda\int d^2\theta\ \Phi\Phi'\Phi'' \to \lambda[f_L^T\sigma_2 f_L' h'' + f_L'^T\sigma_2 f_L'' h + f_L''^T\sigma_2 f_L h']$$

$$+ \lambda^2[h^*hh^{*\prime}h' + h^*hh^{*\prime\prime}h'' + h^{*\prime}h'h^{*\prime\prime}h''],$$

where θ are left-handed Grassman variables, m is a mass and λ is a dimensionless coupling constant. In the above we have eliminated the F terms through their equations of motion. Higher polynomials in Φ lead to nonrenormalizable interactions and can only appear in the effective action. We have omitted to write the supersymmetric invariant that contains only one field because we will deal with superfields with gauge quantum numbers. Also, we have not written out the part of L that contains the kinetic terms and the gauge interactions. Of course, the complex conjugates must be added to the above to respect TCP.

The minimal SU_5 theory contains 3 $(\bar{5}_f + 10_f)$, 5_H, and 24_H. The subscript f and H stand for fermion and Higgs, respectively. In order to supersymmetrize it, we need therefore 3 $(\Phi_{\bar{5}} + \Phi_{10})$, a superfield H_5 that contains the Higgs field 5_H, another superfield $H_{\bar{5}}$ to cancel the chiral anomaly created by the Weyl spinor in H_5, and a superfield H_{24} that contains 24_H.

Then the most general SU_5 and supersymmetric invariant couplings are given by (here i,j,k are family indices)

$$a_{ij}\ \Phi_{\bar{5}}^i\ H_5\ \Phi_{10}^j + b_{ij}\ \Phi_{10}^i\ \Phi_{10}^j\ H_5 + m\ H_{\bar{5}}\ H_5 + c\ H_{\bar{5}}\ H_{24}\ H_5$$

$$+ M\ H_{24}\ H_{24} + d\ H_{24}\ H_{24}\ H_{24}\ ,$$

which generalize the minimal SU_5 couplings, together with

$$\alpha_{ijk}\,\Phi^{i}_{\bar{5}}\,\Phi^{j}_{\bar{5}}\,\Phi^{k}_{\bar{5}} + \mu_i\,\Phi^{i}_{\bar{5}}\,H_5 + \lambda_i\,\Phi^{i}_{\bar{5}}\,H_5\,H_{24} \quad ,$$

which are new couplings not present in the minimal SU_5. In the following we wish to examine the effect of these couplings in GUTs. For future reference, call the first group of couplings I, the second group II.

In the standard supersymmetric GUTs[12] one finds only group I couplings. They include only the minimal SU_5 couplings and thus respect a global phase symmetry which, after electroweak breaking, becomes B-L. Thus, as in the nonsupersymmetric case, neutrinos are massless. The gauge hierarchy problem is handled as follows: when 24_H gets its vacuum value, V, it generates different masses for the Higgs doublets and color triplets given by m - 3cV and m + 2cV, respectively. The mass of the Higgs doublet is set to zero by choosing m = 3cV, while the Higgs triplet gets a large mass. In the absence of supersymmetry breaking electroweak breaking would never occur. Supersymmetry must be broken, and in a way which can be arranged to induce a tiny negative mass squared for the Higgs doublet (thus "solving" the gauge hierarchy problem?). Here we do not concern ourselves with the particular mechanisms of supersymmetry breaking - we just assume it can be done in the required way.[13, 14]

In SU_5, the simplest possible proton decay modes occur via the effective interactions[15] $\bar{5}_f 10_f \overline{10}_{\bar{f}} 5_{\bar{f}}$, $10_f \overline{10}_{\bar{f}} 10_f \overline{10}_{\bar{f}}$, and $\bar{5}_f 10_f 10_f 10_f$, where the subscripts f and $\bar{f}$ label fields that transform as (1,2) and (2,1) respectively under the Lorentz group $SU_2 \times SU_2$. The first two lead to both family-skew and family-diagonal decay modes; the last one only to family-skew decays, since one needs at least two 10_f in order to make an SU_5 invariant. One can estimate the strength of these interactions by cataloging the quantum numbers appearing in the s-, t-, and u- channels of these amplitudes. If they match with a field appearing in the theory that couples to both in and out two particle states, the resulting diagram will appear at tree level; if they match with a two particle state, the dominant diagram will occur at one loop, etc. Thus, the first two interactions can have a vector [~ (2,2) under the Lorentz group] appearing in one channel since $(\bar{5},1,2) \times (5,2,1)$ and $(10,1,2) \times (\overline{10},2,1)$ both contain (24,2,2) which are the SU_5 gauge bosons. Hence, vector exchange contributes at tree level to both, as is well known. By looking at the $\bar{5} \times 10$ and 10×10 channel we see that we can also have scalar exchange at tree level. These correspond to Higgs triplets contribution to the first two interactions. The third possible invariant, $\bar{5}_f 10_f 10_f 10_f$, cannot be caused by an odd number of vector exchanges, but can be generated by Higgs exchange at tree level.

To see this, note that

$$(\bar{5},1,2)\ (10,1,2) = (5 + 45,\ 1,\ 1 + 3) \quad ,$$

$$(10,1,2)\ (10,1,2)_A = (\bar{5} + \overline{45},1,1) + (\overline{45},1,3),$$

so that the two particle channels have (5,1,1) - the quantum number of a Higgs quintet in common. In theories where the Higgs coupling to $\bar{5}\cdot 10$ is not the same as the one coupling $10\cdot 10$ this interaction does not get any tree level contributions. Examples of such theories are those with Peccei-Quinn symmetry and/or supersymmetry. [Anyhow more than one Higgs is needed to generate baryon number asymmetry.] Thus the lowest order contribution to this diagram will be given by at least a one loop diagram.

In the standard SU_5-supersymmetric theory, the Grand Unification scale is pushed up with the result that gauge boson contributions to the first two interactions give lifetimes much longer than the present limit, and the same is true for the Higgs triplet contributions, so that all tree level diagrams are suppressed. [Can their dominant contribution come from one loop diagrams?] However, there are many more fermions which appear as the supersymmetric partners of the minimal SU_5 set. These are the partners of the Higgs, transforming as $(\bar{5},1,2)$ and $(5,1,2)$ and the "gauginos" transforming as $(24,1,2) + (24,2,1)$. These gauginos lead to new tree-level four fermion interactions of the form $\bar{5}_f 10_f 24_f \bar{5}_f$ and $10_f 10_f 24_f 5_f$, which are capable of causing proton decay at the one loop level, by contributing to the family-skew effective interaction $\bar{5}_f 10_f 10_f 10_f$. This is apparently the dominating mechanism for proton decay in standard supersymmetric theories.[16] Incidentally, it should be noted that there are also new tree level interactions of the form $5_{\bar{f}} \bar{5}_f 24_f 24_f$ and $10_f \overline{10}_{\bar{f}} 24_f 24_f$ which give one-loop contributions to the standard family-diagonal couplings $\bar{5}_f 10_f 5_{\bar{f}} \overline{10}_{\bar{f}}$ and $10_f \overline{10}_{\bar{f}} 10_f \overline{10}_{\bar{f}}$, with strength dependent on the gaugino masses.

Let us now discuss the effects of group II couplings. First of all, since these couplings break B-L, they will generate neutrino masses.[17] Secondly, the α-type couplings can generate proton decay interactions by means of a spinless color triplet exchange. However, this spinless particle is, in the absence of mixing angle, the supersymmetric partner of ordinary quarks; it cannot be made very massive, otherwise supersymmetry would be of no help for the gauge hierarchy. Thus, proton decay would occur at tree level at a very rapid rate. This bad situation can be remedied only by finding ways in which the group II couplings can be made very small. This is the reason why these group II couplings have been traditionally ignored - they are too dangerous. Our point is that these couplings should not be ignored in a theory which already contains very small ratios (to satisfy the gauge hierarchy). Furthermore, these new couplings do not merely provide small corrections to the already present group I couplings; rather, they have very definite signatures: neutrino masses and family-diagonal proton decay. Also, it gives the hope of relating neutrino masses to certain types of proton decays.

We have not yet fully analyzed the whole situation, but we wish to present the effect of some of the group II coupling by presenting a one-family model. There the α-type couplings do not exist and we only have the μ and λ terms. The relevant part of the superpotential is

$$a\, \Phi_{\bar{5}}\, H_{\bar{5}}\, \Phi_{10} + b\, \Phi_{10}\, \Phi_{10}\, H_5 + m\, H_{\bar{5}}\, H_5 + c\, H_{\bar{5}}\, H_{24}\, H_5$$

$$+ \mu\, \Phi_{\bar{5}}\, H_5 + \lambda\, \Phi_{\bar{5}}\, H_{24}\, H_5 + M\, H_{24}\, H_{24} + d\, H_{24}\, H_{24}\, H_{24}.$$

As a first step, one diagonalizes the mass term $(m\, H_{\bar{5}} + \mu\, \Phi_{\bar{5}})\, H_5$, by mixing $H_{\bar{5}}$ and $\Phi_{\bar{5}}$. If we introduce the angles α, β, γ such that

$$\tan\beta = \frac{\mu}{m};\ \tan\gamma = \frac{\lambda}{c};\ \alpha = \beta - \gamma \quad ,$$

we find that the superpotential becomes

$$a \cos\beta\, \Phi'_{\bar{5}}\, H'_{\bar{5}}\, \Phi_{10} + m'\, H'_{\bar{5}}\, H_5 + c'\, \{\cos\alpha\, H'_{\bar{5}} + \sin\alpha\, \Phi'_{\bar{5}}\}\, H_5 H_{24}$$

with the other terms unchanged, and where $m' = (m^2 + \mu^2)^{1/2}$, $c' = (c^2 + \lambda^2)^{1/2}$, and

$$H'_{\bar{5}} = \cos\beta\, H_{\bar{5}} + \sin\beta\, \Phi_{\bar{5}};\ \Phi'_{\bar{5}} = -\sin\beta\, H_{\bar{5}} + \cos\beta\, \Phi_{\bar{5}}.$$

At this stage the only difference with the standard supersymmetric model is the appearance of the $\sin\alpha$ coupling. Next, the spinless field of H_{24} gets a vacuum value, breaking SU_5, but not supersymmetry

$$h_{24} = V \begin{pmatrix} 2 & & & & \\ & 2 & & & \\ & & 2 & & \\ & & & -3 & \\ & & & & -3 \end{pmatrix} \quad .$$

The superfields are now labeled by their $SU_2 \times SU_3^c$ quantum numbers. The new fields

$$\hat{H}_{(2,1)} = \cos\delta\, H'_{(2,1)} - \sin\delta\, \Phi'_{(2,1)};\ \tilde{\Phi}_{(2,1)} = \sin\delta\, H'_{(2,1)} + \cos\delta\, \Phi'_{(2,1)},$$

where

$$\tan\delta = \frac{3Vc' \sin\alpha}{m' - 3c'V \cos\alpha},$$

are the mass eigenstates in the doublet sector. The $\tilde{\Phi}_{(2,1)}$ superfield is massless and the $\tilde{H}_{(2,1)}$ superfield has a mass

$$m_2^2 = (m' - 3Vc' \cos\alpha)^2 + (3Vc' \sin\alpha)^2.$$

In the standard model only the first term was present.[18] Hence, m_2 can never be set equal to zero, unless $\alpha = 0$. Gauge hierarchy requires only that m_2 be small compared to V, so that it may be overtaken by negative contributions from fermion loops. Thus, the imposition of a gauge hierarchy demands that $\sin\alpha$ be very small. [The smallness of α may be understood when supergravity is included.] In one approach, m_2^2 gets a negative contribution[14] from gauginos, triggering electroweak breaking. Here an extra contribution may not be unwelcome since in supergravity, gravitino exchange will also contribute negatively to m_2^2. In the color triplet sector, the mixing is described by another angle ε. Introduce the mass eigenstate superfields.

$$\tilde{H}_{(1,\bar{3})} = \cos\varepsilon\, H'_{(1,\bar{3})} - \sin\varepsilon\, \Phi'_{(1,\bar{3})}; \quad \tilde{\Phi}_{(1,\bar{3})} = \sin\varepsilon\, H'_{(1,\bar{3})} + \cos\varepsilon\, \Phi'_{(1,\bar{3})}$$

with

$$\tan\varepsilon = \frac{2c'V \sin\alpha}{m' + 2c'V \cos\alpha}.$$

The superfield $\tilde{\Phi}_{(1,\bar{3})}$ is massless and $\tilde{H}_{(1,\bar{3})}$ has a mass

$$m_3^2 = (m' + 2c'V \cos\alpha)^2 + (2c'V \sin\alpha)^2.$$

If one chooses the input parameters so as to make m_2 very small,

m_3 gets a value of the order of V, the superGUT scale. The part of the superpotential that contains the massless or nearly massless superfield now reads

$$a \cos 2\beta \{\cos 2\delta\ \tilde{\Phi}_{(2,1)}\ \tilde{H}_{(2,1)}\ \Phi_{(1,1)} + \cos 2\varepsilon\ \tilde{\Phi}_{(2,1)}\ \tilde{\Phi}_{(1,\bar{3})}\ \Phi_{(1,\bar{3})}$$

$$- \sin(\delta + \varepsilon)\ [\tilde{H}_{(2,1)}\ \tilde{H}_{(1,\bar{3})} + \tilde{\Phi}_{(2,1)}\ \tilde{\Phi}_{(1,\bar{3})}]\ \Phi_{(2,3)}$$

$$- \cos(\delta + \varepsilon)\ [\tilde{H}_{(2,1)}\ \tilde{\Phi}_{(1,\bar{3})} - \tilde{H}_{(1,\bar{3})}\ \tilde{\Phi}_{(2,1)}]\ \Phi_{(2,3)}\}$$

$$+ 2b\ \Phi_{(1,\bar{3})}\ \Phi_{(2,3)}\ H_{(2,1)}\ .$$

In the standard model ($\alpha = 0$, or $\delta = \varepsilon = 0$), the term proportional to $\sin(\delta + \varepsilon)$ would have been absent. In this embryonic model, its effect is to create a new type of leptoquark interaction of the form $\tilde{\Phi}_{(1,\bar{3})}\ \tilde{\Phi}_{(1,3)}\ \Phi_{(2,3)}$. We should emphasize that the dangerous spinless field in $\tilde{\Phi}_{(1,\bar{3})}$ can act here only as a spinless leptoquark because we have only one family; in a model with several families, it would act as an anti-quark as well, thus leading directly to proton decay. These decays can be family-diagonal, and their strength is tied to the amount of B-L violation, i.e. neutrino masses. Since it cannot be very massive, we must rely on the smallness of $\sin(\delta + \varepsilon)$ to suppress the proton decay rate, linking it to small neutrino masses. Still, from this simple model we have learned that the group II couplings are limited in size by the gauge hierarchy!

Let us conclude this presentation by mentioning the possible role of the family group in N=1 supersymmetry. We have learned that in such theories there appear two superfields which transform as $\overline{5}$ of SU_5. It may therefore be natural to introduce a new SU_2 which mixes them. However, this SU_2 must be badly broken. In the case of many families, it can be used to prevent the dangerous α-type couplings. For instance, one might consider an $SU_5 \times SU_2$ model where the superfield content is $n(\bar{5},2) + n(10,1) + n(5,1)$, where n is the number of families. If we write the n doublets $(\bar{5},2)$ as $(\Phi^i_{\bar{5}}, H^i_{\bar{5}})$, then SU_2 invariance only allows couplings of the form $\Phi^i_{\bar{5}} H^i_{\bar{5}} \Phi^k_{10}$.

To conclude, let us emphasize that the minimal supersymmetric Grand Unified theory contains B-L violation, as well as the possibility of having dominant family symmetric proton decay modes. However, these interactions are limited in strength, lest they upset the usual gauge hierarchy. It is amusing that these new processes do not contribute small correction to processes already existing in standard superGUT, but rather predict new physical phenomena with distinctive signatures, such as neutrino masses and family-diagonal proton decay. We hope to come back to these problems elsewhere.[19]

Acknowledgements

I thank Prof. Sikivie, Dr. M. Chase and M. Bowick for useful discussions.

References

1. R.D. Peccei and H. Quinn, Phys. Rev. Lett. 38, 1440 (1977).
2. J. Ellis and M.K. Gaillard, Nucl. Phys. B150, 141 (1979).
3. S. Weinberg, Phys. Rev. Lett. 40, 223 (1978); F. Wilczek, Phys. Rev. Lett. 40, 279 (1978).
4. L. Abbott and P. Sikivie, Brandeis preprint, Sept. 1982; J. Preskill, M. Wise and F. Wilczek, Harvard preprint, Sept. 1982; M. Dine and W. Fischler, Penn. preprint, Sept. 1982.
5. D.A. Dicus, E.W. Kolb, V.L. Teplitz and R.V. Wagoner, Phys. Rev. D22, 839 (1980); M. Fukugita, S. Watamura and M. Yoshimura, Phys. Rev. Lett. 48, 1522 (1982).
6. For a review see R.D. Peccei in Neutrino 81, Hawaii, July 1981.
7. J. Ipser and P. Sikivie, University of Florida preprint TP-83-1, January 1983.
8. A.H. Guth, Phys. Rev. D23, 347 (1981).

9. E. Witten, Nucl. Phys. 177B, 477 (1981).

10. A. Albrecht and P.J. Steinhardt, Phys. Rev. Lett. 48, 1220 (1982), A.D. Linde, Phys. Lett. 108B, 389 (1982).

11. For a review see "Supersymmetry and Supergravity" by J. Wess, lectures given at Princeton University, 1982, to be published by Princeton University Press.

12. L. Maiani, in Proc. Ecole d'Ete de Physique des particules (Gif-sur-Yvette, 1979), p. 3.
E. Witten, Nucl. Phys. B186, 513 (1981).
S. Dimopoulos and H. Georgi, Nucl. Phys. B193, 150 (1981).
N. Sakai, Z. Phys. C11, 153 (1982).
H.P. Nilles and S. Raby, Nucl. Phys. B198, 102 (1982).

13. S. Dimopoulos and S. Raby, Nucl. Phys. B192, 353 (1982).
M. Dine, W. Fischler, and M. Srednicki, Nucl. Phys. B189, 575 (1981).

14. J. Ellis, L.E. Ibañez and G.G. Ross, Phys. Lett. 113B, 283 (1982).

15. S. Weinberg, Phys. Rev. Lett. 43, 1566 (1979); F. Wilczek and A. Zee, Phys. Rev. Lett. 43, 1571 (1979).

16. S. Weinberg, Supersymmetry at ordinary energies - HUTP-81/A047; N. Sakai and T. Yanagida, Proton decay models, Max Planck preprint (Oct, 1981); S. Dimopoulos, S. Raby, and F. Wilczek, Phys. Lett. 112B, 133 (1982).

17. P. Ramond, Sanibel talk, Feb. 1979, CALT-68-709 unpublished. Nonzero neutrino masses are made possible by the presence of extra fields, the supersymmetric partners of the ordinary particles. For the effect of extra fields in generating neutrino masses in non supersymmetric theories, see for instance A. Zee, Phys. Lett. 93B, 389 (1980).

18. For a more sophisticated potential which ensures $m_2 = 0$ naturally, see Ref. (14).

19. M. Bowick, M. Chase and P. Ramond, in preparation.

SPONTANEOUS SUPERSYMMETRY BREAKING AND METASTABLE VACUA

G. Domokos and S. Kovesi-Domokos

Department of Physics, Johns Hopkins University

Baltimore, MD 21218

Abtract

The question of a dynamical breakdown of supersymmetry is discussed and a recently proved "no-go theorem" is reviewed. We analyze the possibility that a supersymmetry-breaking vacuum is metastable. Long lifetimes of such vacua can be easily achieved without fine-tuning the parameters of the models. Models with metastable vacua can be made quite economical: internal symmetries and supersymmetry can be broken by the same Higgs fields at comparable mass scales.

Introduction

The recent revival of interest in sypersymmetric (SUSY) models is due to the fact that SUSY protects mass hierarchies in grand unified theories. This is a well known fact, and it need not be reviewed here. As a quick reminder, we just recall that if SUSY is unbroken, the propagator of a chiral superfield is proportional to a delta function in the Grassmann coordinates, viz.

$$\langle T(\phi(P_1)\phi(P_2))\rangle = \delta(\Theta_1 - \Theta_2) f((y_1 - y_2)^2),$$

where, as usual, $y^\mu = x^\mu + i\Theta\sigma^\mu\bar{\Theta}$ and P_i stands for a point in superspace. The contents of the nonrenormalization theorem[1] is simply that in any diagram containing closed loops (those are the ones endangering a tree level mass hierarchy) one always collects higher powers of $\delta(\Theta_1 - \Theta_2)$; thus, in view of $(\delta(\Theta_1 - \Theta_2))^2 = 0$, the loop contributions vanish.

Beautiful as it is, SUSY (if it has anything to do with Nature) must be broken and, in order to preserve its good properties, it must be broken spontaneously. The trouble is that in global SUSY the vacuum energy is positive semidefinite; hence, if any model admits a sypersymmetric vacuum, that is absolutely stable. It may happen that - in contrast with one's intuition - global SUSY is just an unacceptable approximation and (super) gravity plays an essential role even at distances small compared with cosmological scales: this approach is discussed by Pran Nath at this Conference and we refer the reader to his paper appearing in these Proceedings. We take the conservative approach according to which *global* SUSY should be an acceptable approximation at "small" distance scales. It has been known for about eight years that global SUSY *can be* broken spontaneously[2] at the cost of adding some more Higgs fields to the models in order to prevent a supersymmetric vacuum from developing. Grand unified models of this type were pursued by several authors and they have been concisely reviewed, for instance, in a recent paper by Dine.[3]

It is very tempting to speculate that the large number of Higgs fields required by SUSY GUTs is a hint that those Higgs fields are, in fact, composite ones, presumably, bilinears of the "good" superfields describing quarks and leptons. Thus, SUSY would be broken by bilinear condensates[4], very much as chiral symmetry is believed to be broken ("dynamical" SUSY breaking). This would be very nice: the (many) parameters in the effective Higgs potential would be calculable from the dynamics. Unfortunately, such a scenario is an overly optimistic one: we now turn to discuss that subject.

Absence of stable bilinear condensates in SUSY theories

Attempts at finding stable bilinear condensates in SUSY

theories (by means of some nonperturbative calculation) failed[5]: either there appeared no condensate at all, or it was accompanied by a supersymmetric solution to the dynamical equations, hence the condensate was unstable. The question is whether this is a general feature of SUSY models or we were just not clever enough to devise a suitable model and/or a successful approximation scheme. The answer is that (in four Bosonic dimensions) *no stable bilinear condensate of chiral superfields can exist.*[6] Instead of going through the somewhat tedious manipulation leading to this result outlined in ref. 6, let us try to understand in an intuitive way how such a result emerges.

Let us recall that the emergence of a condensate of both elementary and composite fields is best studied in the framework of an effective action formalism[7], which, in turn, can be generalized to SUSY theories.[8] However, the structure of the effective actions for elementary and composite fields is quite different. For an elementary field, $\Phi(x)$, denoting its vacuum expectation value (VEV) by $\langle\Phi(x)\rangle \equiv \phi$, the effective potential contains a tree contribution plus corrections due to vacuum loops. (There is no kinetic energy contribution to the effective action, because one is looking for a translationally invariant ϕ.)

By contrast, a local composite field can only be defined by means of a rather nasty limiting procedure, thus one has to study the VEV, $G(P_1,P_2) \equiv \langle T(\phi(P_1)\phi(P_2))\rangle$, i.e. the two-point function. Its effective action is given by the Cornwall-Jackiw-Tomboulis formula[9], viz.

$$S[G] = \mathrm{Tr}\{G_0^{-1}G - 1 - \ell n GG_0^{-1}\} - V_2[G], \tag{2.1}$$

where $V_2[G]$ stands for the sum of the two-particle irreducible vacuum-to-vacuum diagrams. The trace is understood in the functional sense (summation over discrete indices, integration over continuous variables); G_0 stands for the noninteracting two-point function. The main result of ref. 8 is that, once a SUSY theory is formulated in terms of superfields over superspace, effective actions of various kinds are exactly of the same form as for a non-SUSY theory. The only change is that now "T_r" also includes integration over the Grassman variables and G_0 in (2.1) is, by definition, sypersymmetric. (In other words, the structure of the effective action is insensitive to the nature of the base manifold over which the theory is defined.) In fact, this is all there is to the "no-go" theorem, the rest is "supersymmetric technology". For the sake of simplicity, let us illustrate how these observations lead to the absence of stable pair condensates in the case of composite fields of the F-type (products of only left-handed or only right-handed chiral fields); the analysis is slightly more complicated in the general case, see ref. 6.

One can decompose G into sypersymmetric and SUSY-breaking parts, viz.

$$\begin{aligned} G(P_1,P_2) &= \delta(\Theta_1 - \Theta_2)G_S(y_1 - y_2) \\ &+ \varepsilon(G_1(y_1 - y_2) + (\Theta_1 \cdot \Theta_2)G_2(y_1 - y_2) \\ &+ \Theta_1^2\Theta_2^2\, G_3(y_1 - y_2)), \end{aligned} \tag{2.2}$$

cf. ref. 6. The parameter ε was only introduced for purposes of keeping track of SUSY-breaking pieces of the two-point functions. Variation of (2.1) with respect to G gives the Dyson-Schwinger equation, viz.

$$G_0^{-1} - G^{-1} - \frac{\delta V_2}{\delta G} = 0, \tag{2.3}$$

the last term being the proper self-energy. The inverse of G possesses a decomposition analogous to (2.2), say

$$\begin{aligned} G^{-1}(P_1,P_2) &= \delta(\Theta_1 - \Theta_2)X_s(y_1 - y_2,\varepsilon) \\ &+ \varepsilon(X_1(y_1 - y_2,\varepsilon) + (\Theta_1 \cdot \Theta_2)X_2(y_1 - y_2,\varepsilon) \\ &+ \Theta_1^2\Theta_2^2X_3(y_1 - y_2,\varepsilon)). \end{aligned} \tag{2.4}$$

The important observation here is that the coefficients X_i (i = s, 1, 2, 3) are smooth functions of ε in a sufficiently large domain around $\varepsilon = 0$: this follows from SUSY kinematics.[6] The self-energy part is a smooth functional of G, essentially because the dynamical SUSY breaking is expected to be generated by a bilinear condensate and not, say, by cubic, quartic, etc. condensates. Moreover $V_2 \to 0$ in the limit $\varepsilon \to 0$: this is just the nonrenormalization theorem combined with the decomposition (2.4). (This is somewhat stronger than a perturbative result: G may be any function of the coupling constants.) It follows that $\varepsilon = 0$, $G = G_0$ is always a possible solution of the Dyson-Schwinger equation, hence any SUSY-breaking solution is unstable due to the fact that SUSY vacuum has the lowest energy.

This result is of the same type as in non-SUSY theories (cf. the model of Nambu and Jona-Lasinio[10], the BCS theory of superconductivity, etc.). All theories leading to the formation of pair condensates have symmetric solutions too; however, because of the special properties of global SUSY theories (the energy of the vacuum is an order parameter), the stability situation is different. Our result establishes the absence of stable pair condensates of chiral superfields; however, it says nothing about the possible pair condensation triggered by condensates of gauge superfields ("instanton effects"). This is an interesting and rather ill-understood effect in SUSY theories: it may happen that such a mechanism will be ultimately found responsible for the breakdown of SUSY. In what follows, however, we explore another, rather amusing, alternative.

Metastable vacua?

Let us recall how the O'Raifertaigh mechanism[2] (and variations thereon) work. Consider the Higgs sector of a SUSY theory, ϕ being a Higgs superfield. (Internal indices are suppressed for the sake of simplicity.) The superpotential is written in the form $W = U(\Phi)\delta(\bar{\Theta})$ + h.c., so that the field equations for a translationally invariant VEV ($\delta\phi/\partial x^\mu = 0$) read[8]:

$$\int d^2\,\Theta\phi(\Theta) + \frac{\partial U^+}{\partial \phi^+} = 0. \qquad (3.1)$$

This equation is still a function of Θ, but it is obvious that the F-component of ϕ is driven by the A-component[11] of the "superforce", $\partial U^+/\partial\phi^+$. The absence of a SUSY solution is guaranteed if the driving term cannot vanish: this then leads, in a well known way, to the proliferation of Higgs superfields.[2]

However, do we really have to exclude the existence of SUSY solutions? Our Universe today would look very much the same as we know it if, in the course of its evolution, it somehow landed in a "false" (metastable) SUSY-breaking vacuum, having a sufficiently long lifetime, for instance, $\Gamma/V < 10^{-116}\,yr^{-1}\,cm^{-3}$, where V is the volume of the Universe. (Today, $V = V_0 \cong 10^{84}\,cm^3$.) This question was recently investigated by Turner and Wilczek in the framework of ordinary grand unified models[12] without SUSY. They point out that we can easily live in a false vacuum. (Moreover, should the false vacuum decay into the true one, we would not have a long time to contemplate what happened to us since the initially formed bubble of the true vacuum expands at the speed of light; so we do not have to worry anyway...)

The interesting feature of such a scenario is that, we do not need more Higgs fields (elementary or composite) beyond the ones required by the breaking of internal symmetries, thus such models can be made quite economical. In order to understand the problems and possibilities involved, let us study a toy-model in which, in essence, we ignore complications due to internal symmetries. The only way in which we mimick the presence of some internal symmetry (presumably a simple group) is that we deny the existence of terms linear in the Higgs superfield in the superpotential. It is then immediately clear that a metastable SUSY-breaking vacuum cannot arise in any model which contains chiral superfields only and which is a renormalizable one (Wess-Zumino models).

This can be seen as follows. A metastable, SUSY-breaking vacuum must be separated from the SUSY vacuum by a barrier. If the superpotential is of the form

$$W = U(\Phi)\delta(\bar{\Theta}) + h.c., \tag{3.2}$$

with U at the most cubic in ϕ (renormalizability), the equation for the F-component reads (cf. eq. (3.1)):

$$F + f^*(A^*) = 0, \tag{3.3}$$

$$\Phi(A) \equiv \int d^2\Theta\delta(\Theta)\frac{\partial U}{\partial \phi}$$

is quadratic in A*. The scalar potential is given by

$$V = |f(A)|^2 \quad , \tag{3.4}$$

and its extrema are located where

$$f^* \frac{\partial f}{\partial A} = f\frac{\partial f^*}{\partial A^*} = 0 \quad , \tag{3.5}$$

i.e. either f =0 (which is the SUSY solution) or $\partial f/\partial A = 0$, which could break SUSY. However, $\partial f/\partial A$ is a linear function of A, it has one root, whereas the existence of a barrier would require f to have two extrema, a maximum and a minimum.

We thus try an effective superpotential of the form:

$$U(\Phi) = \frac{m}{2}\Phi^2 - \frac{g}{3}\Phi^3 + \frac{h}{4m}\Phi^4 \quad , \tag{3.6}$$

where m is some mass scale and g, h are dimensionless coupling constants. (For the sake of definiteness, we take $m > 0$, $g > 0$, $h > 0$.) Taken literally, this leads to a nonrenormalizable theory; however, SUSY still helps somewhat, because all vertices are of the F-type. Therefore, loop contributions to amplitudes containing only Φ's or only Φ^+'s as external legs still vanish: in particular, the coefficients of U do not pick up radiative corrections. (The horror of nonrenormalizability manifests itself only in amplitudes with both Φ and Φ^+ external legs.) We do not discuss how such an effective potential arises; presumably, it does through the coupling of the Higgs field to gauge superfields (or to supergravity?).

We do believe, however, that the precise form of U does not affect our main conclusions as long as its qualitative behavior is similar to (3.6) in the relevant region of the space of the Φ and Φ^+.

We now have in our toy-model:

$$f(A) = mA - gA^2 + \frac{h}{m} A^3 \quad . \tag{3.7}$$

It will be convenient to measure A in units of m; we put

$$A \equiv \frac{m}{g} z, \qquad f(A) \equiv \frac{m^2}{g} H(z) \quad .$$

Thus $H(z) = z(1 - z + z^2\xi/3)$, where $\xi = 3h/g^2$. It is easily seen that H(z) has only one real root (z = 0) and two extrema if ξ is in the range,

$$\frac{3}{4} < \xi < 1 \tag{3.8}$$

In this range of ξ, the scalar potential, $V = m^4 g^{-2} |H(z)|^2$ has two minima located at

$$z = 0 \text{ and } z = z_+ = \frac{1}{\xi}(1 + \sqrt{1 - \xi}) \quad , \tag{3.9}$$

separated by a maximum at $z = z_- = \xi^{-1}(1 - (1-\xi)^{\frac{1}{2}})$. Thus the SUSY vacuum (z=0, f(0)=0) and the symmetry breaking one are indeed separated by a barrier. This time, however, the minimum at z_+ breaks both SUSY and whatever internal symmetry we think we put into the model. In the range (3.8) of the parameter ξ both vacuum expectation values, A and, say, F/m, are roughly of the same order of magnitude. Indeed, as ξ runs through its allowed range, z_+ and $H(z_+)$ vary between the values

$$1 < z_+ < 2, \quad \frac{1}{3} < H(z_+) < \frac{117}{128} \; .$$

Our readily checks that at z_+, the Fermi component of ϕ becomes a "Goldstino", as expected. This follows simply from the elementary formula,

$$\int d^2\Theta\Phi^n = n[A^{n-1}F - \frac{1}{2}(n-1)A^{n-2}(\psi\psi)], \tag{3.10}$$

showing that the mass of the Fermi component is proportional to $\partial f/\partial A$. Decoupling then follows from the SUSY Ward identities.[8]

How stable is the false vacuum, or rather, how hard do we have to work in order to make it sufficiently long-lived? The semiclassical formula for the decay rate of the false vacuum is given in the papers of Coleman and Callan and Coleman[13]:

$$\Gamma/V = D\,\exp(-\,S_E), \tag{3.11}$$

where S_E is the extremal value of the Euclidean action of the scalar field and D is the ratio of two Van Vleck-Morette determinants; on the basis of dimensional arguments we expect D to be of the order of m^4. In our toy-model, the classical field configuration extremizing S_E is real. For an SO(4) symmetric bubble one finds with $x = m(\tau^2 + \underline{x}^2)^{1/2}$:

$$S_E = \frac{2\pi^2}{g^2}\int_0^\infty x^3 dx\,(z'^2 + H(z)^2). \tag{3.10}$$

Although the form of the explicit solution of $\delta S_E/\delta z = 0$ is more complicated here than in a simple scalar field theory with a quartic interaction, the qualitative properties are easy to analyze; the integral in (3.10) is expected to be order unity (probably somewhat larger), because the dimensionless field z tunnels down smoothly from $z(0) = z_+ \simeq 1$ to zero and the integral is rapidly converging at $x = \infty$. (The classical solution behaves as exp(-x), hence the integrand in (3.10) goes as exp(-2x)). As a rough guess, we put $D^{1/4} \simeq m \simeq 1$ TeV (global SUSY is expected to break down around such an effective mass scale), $S_E \simeq 1$; then, in order to achieve $\Gamma \simeq 10^{31}$ yrs "today" ($V \simeq V_0$), one needs $g \simeq 0.2$ No "fine tuning" of the theory is necessary, nor does one need fantastically large or unreasonably small values of the coupling constants: given g, the constant H is constrained by (3.8).

The basic question is whether the Universe can land with some reasonable probability in the false vacuum as it cools down after the Big Bang. We are in the process of analyzing this question; however, there is already some indication that this may be the case.

It was shown by Das and Kaku and by Girardello, Grisaru and Salomonson that SUSY is broken at finite temperatures.[14] Clearly, what one has to determine is whether fluctuations around the true or the false vacuum are more strongly coupled: this, in turn, determines (at least in a high-temperature expansion) the temperature dependence of the free energy. As it turns out, the scalar coupling constants (V^{III}, V^{IV}, ...) are smaller at the "false" vacuum; thus our guess is that at high temperatures ($T \gg m$), the free energy of the false vacuum is smaller: the Universe seems to have a chance of getting hung up in the SUSY breaking vacuum.

Conclusion

Supersymmetry appears to be unique among all the symmetries discovered (invented?) by mankind: it has never been unbroken during the history of the Universe. Right after the Big Bang, it was broken by finite temperature effects, today (at essentially zero temperature) it must be broken by vacuum expectation values of some Higgs fields, whether those are elementary or composite. We do know that if these Higgs fields are composite, the structure of the theory must be quite rich, for a SUSY-breaking pair condensate in any simple renormalizable theory would have dissolved long ago: this is the contents of the "no-go" theorem we discussed.

Nevertheless, SUSY may well be the Fairy Godmother protecting our gauge hierarchy from being destroyed (or, in plain English, our protons from decaying too rapidly[15]). Unless, of course, we live in a metastable world as discussed: Fairy Godmother may be just waiting to turn suddenly into Wicked Witch, making our neutrinos, electrons, quarks,... weigh the same, with a sudden energy release of quite a few TeV/cm^3 in the process. We shall certainly not live to see whether (or if) that happens and how the world will look like afterwards.

Acknowledgements

We wish to thank the Organizers of the Orbis Scientiae

Conferences for the opportunity of presenting our view of supersymmetry in the stimulating atmosphere of this Conference. This research was supported in part by the U.S. Department of Energy, under Contract No. DE-AC02-76ER03285.

REFERENCES

1. M. Grisaru, W. Siegel and M. Rocek, Nucl. Phys. B159, 429 (1979).
2. L. O'Raifertaigh , Nucl. Phys. B96, 331 (1975).
3. M. Dine, in Proc. Sixth Johns Hopkins Workshop on Current Problems in Particle Theory. (G. Domokos and S. Kövesi-Domokos, Editors). Johns Hopkins University, Baltimore, MD. 1982.
4. S. Dimopoulos and S. Raby, Nucl. Phys. B192, 353 (1981). E. Witten, Nucl. Phys. B185, 513 (1981). M. Dine, W. Fischler and M. Srednicki, Nucl. Phys. B189, 575 (1981).
5. See for instance H. P. Nilles, Phys. Lett. 112B, 409 (1982). G. Domokos and S. Kövesi-Domokos, Phys. Rev. D (to be published).
6. G. Domokos and S. Kövesi-Domokos, Phys. Letters 120B, 101 (1983).
7. G. Jona-Lasinio, Nuovo Cim. 34, 1970 (1964). G. Domokos and P. Surányi, Sov. Jour. Nucl. Phys. 2, 361 (1966). B. Zumino, in Brandeis Lectures in Elementary Particles and Quantum Field Theory. (S. Deser and H. Pendleton, Editors). MIT Press, Cambridge, MA. (1970).
8. G. Domokos and S. Kövesi-Domokos, Johns Hopkins University preprint, JHU-HET 8203 (1982). To be published in the Festschrift honoring F. Gursey's 60th birthday.
9. J,M. Cornwall, R. Jackiw and E. Tomboulis, Phys, Rev. D10 2428 (1974).
10. Y. Nambu and G. Jona-Lasinio, Phys. Rev. 122, 345 (1961).
11. We use the standard decomposition of chiral superfields, viz. $\Phi(y,\Theta) = A(y) + 2^{1/2}(\Theta\cdot\psi(y)) + \Theta^2 F(y)$, cf. J. Bagger and J. Wess, Supersymmetry and Supergravity, Princeton University Press (to be published).
12. M.S. Turner and F. Wilczek, Nature 298, 633 (1982).

13. S. Coleman, Phys. Rev. D15, 2929 (1977). G. Callan and S. Coleman, ibid. D16, 1762 (1977).
14. A. Das and M. Kaku, Phys. Rev. D18, 4540 (1978). L. Girardello, M.T. Grisaru and P. Salomonson, Nucl. Phys. B178, 331 (1981).
15. Cf. IMB Collaboration, talks of F. Reines and M. Goldhaber, these Proceedings.

SUPERGRAVITY GRAND UNIFICATION*

Pran Nath

Northeastern University

Boston, MA 02115

and

R. Arnowitt†

Harvard University

Cambridge, MA 02138

and

A. H. Chamseddine

Northeastern University

Boston, MA 02115

(Presented by Pran Nath)

Abstract

A review of the recent proposal of Supergravity Grand Unification is given. Topics include the structure of Supergravity GUTS, symmetry breaking through Supergravity induced effects, protection at low energy from intermediate and superheavy mass scales, effective potential and the particle spectrum at low energy.

*Research is supported in part by the National Science Foundation under Grant No. PHY 77-22864 and Grant No. 80-8333.

†On sabbatical leave from the Department of Physics, Northeastern University, Boston, MA 02115.

Experimental consequences of Supergravity GUTS and in particular the decays of W and Z into photino, Wino and Zino modes and their branching ratios in various channels are also discussed.

I. Introduction

Recently a new class of unified gauge theories were introduced[1] which possess the gauge invariance

$$G \times (N = 1 \text{ Supergravity}) , \tag{1.1}$$

where G = SU(5), SO(10), etc. Considerable progress has occurred over the past few months in the investigation and exploration of the physical implications of these ideas further.[2-13] We summarize some of the main features of these developments in this talk. The paper is structured with the following content in the various sections.

Sec. II. Structure of Supergravity GUTS

Sec. III. Gravity Induced Symmetry Breaking

Sec. IV. Gauge Hierarchy and Low Energy Theory

Sec. V. Low Energy Particle Spectrum

Sec. VI. Experimental Consequences.

II. Structure of Supergravity GUTS

The basic multiplets of supergravity grand unified theories (Supergravity GUTS) are N = 1 matter multiplets and the N = 1 supergravity multiplet. As in globally supersymmetric grand unified theories (SUSY GUTS),[14, 15] the matter multiplets consist of a left-handed chiral (F-type) multiplet

$$\Sigma^a = (Z^a, X_L{}^a, h^a) \quad , \tag{2.1}$$

which belongs to a reducible representation of the gauge group G and a vector (D-type) multiplet

$$V = (C, \xi, H, K, V_\mu, \lambda, D) \quad , \tag{2.2}$$

which belongs to the adjoint representation of the gauge group G.

In. Eq. (2.1) a is an internal symmetry index, $Z^a = A^a + iB^a$ are complex scalar fields, X^a_L are left-handed chiral multiplets and $h^a = F^a + iG^a$ are auxiliary (constraint) fields. In Eq. (2.2), C, H, K, D are scalar fields with D also an auxiliary field and ξ and λ are Majorana spinors. The vector multiplet reduces drastically in the Wess-Zumino gauge[16] with only the following elements remaining:

$$(0,\ 0,\ 0,\ V_\mu,\ \lambda,\ D). \tag{2.3}$$

In constructing models, the Higgs mesons and the matter spinors (quarks and leptons) are placed in the chiral multiplets while the gauge vector mesons appear in the vector multiplet. Consequently, supersymmetric theories in general have an array of new objects such as squarks, sleptons, Higgsinos, gauginos, etc. which are super partners of quarks, leptons, Higgs, gauge vector mesons, etc. For N=1 supergravity,[17] the simplest choice is the multiplet with the minimal set of auxiliary fields[18]

$$(e_{a\mu}, \psi_\mu,\ S,\ P,\ A)\quad , \tag{2.4}$$

where $e_{a\mu}$, ψ_μ are the spin 2, spin 3/2 fields and S, P, A_μ are the auxiliary fields of the minimal set.

The coupling to supergravity of a single chiral multiplet $\Sigma = (Z, X_L, h)$ and a single vector multiplet of Eq. (2.2) is of the form[19]

$$L = L_{SG} + L_G + L_F + L_D \quad , \tag{2.5}$$

where L_{SG} and L_G are supergravity and gauge multiplet parts and

$$e^{-1}L_F = \mathrm{Re}[h + uZ + \bar{\psi}_\mu\gamma^\mu X + \bar{\psi}_\mu\sigma^{\mu\nu}\psi_\nu Z] \quad , \tag{2.6}$$

$$e^{-1}L_D = D - \frac{i\kappa}{2}\bar{\psi}_\mu\gamma^5\gamma^\mu\lambda - \frac{2}{3}(SK - PH)$$

$$+ \frac{2}{3}\kappa V_\mu(A_\mu + \frac{3}{8}\, ie^{-1}\varepsilon^{\mu\rho\sigma\tau}\bar{\psi}_\rho\gamma_\tau\psi_\sigma)$$

$$+\frac{4}{3}\, i\kappa\bar{\xi}\gamma_5\sigma^{\mu\nu}D_\mu\psi_\nu - \frac{i\kappa^2}{8}\,\varepsilon^{\mu\nu\rho\sigma}\bar{\psi}_\mu\gamma_\nu\psi_\rho\bar{\xi}\psi_\sigma$$

$$-\frac{2}{3}\,\kappa^2 C\, e^{-1} L_{SG} \quad , \tag{2.7}$$

where $u = S - iP$ and $\kappa = (16\pi G)^{1/2}$, G being the Newtonian constant. Unlike the case of global supersymmetry where only F and D terms are admissible, one finds all elements of the chiral and vector multiplets entering Eqs. (2.6) and (2.7) to guarantee supergravity gauge invariance.

The most general coupling of a single chiral multiplet to supergravity was given by Cremmer et al.[20] and the generalization of these results which couples the N = 1 supergravity multiplet to an arbitrary grand unified gauge group with any number of left-handed chiral multiplets and simultaneously a gauge multiplet was accomplished recently.[1,21,22] The procedure consists of forming the most general F and D multiplets of Eqs. (2.1) and (2.2) which are also invariant under G using the rules of tensor calculus. These multiplets are then coupled to supergravity using the rules of Eqs. (2.6) and (2.7). For this purpose it is found convenient to introduce two functions $g(Z)$ and $\phi(Z,Z^+)$ where $g(Z)$ is the familiar superpotential and $\phi(Z,Z^+)$ represents the lowest element of the most general D-multiplet formed out of the chiral multiplets of Eq. (2.1) maintaining invariance under G. After elimination of the auxiliary fields in the theory and using Weyl scale transformations to guarantee normalized Einstein and Rarita-Schwinger terms in the Lagrangian, one obtains for the Bose part of the Lagrangian the result,[1,21,22]

$$\begin{aligned}
L_B = {} & -(e/2\kappa^2)R(e,\omega) + (3/\kappa^2)G_{,a\bar{b}}[\partial_\mu Z^{b^+} + \tfrac{1}{2}\, ig_e(Z^+V_\mu)^{\bar{b}}] \\
& [\partial^\mu Z^a - \tfrac{1}{2}\, ig_e(V^\mu Z)^a] + (e/\kappa^4)\exp(-G)[3 + (G^{-1})^{\bar{b}a} \\
& G_{,a}G_{,\bar{b}}] - \tfrac{1}{4}\,[T_r(F_{\mu\nu}F^{\mu\nu})] \\
& -\frac{9}{8}\, e(1/\kappa^2\phi)^2[e_\alpha(\phi,a,(T^\alpha Z)_a)]^2 \\
& g_{,b} \equiv \frac{g}{\partial Z^b},\ g_{,\bar{b}} \equiv \frac{\partial g^+}{\partial Z^{b+}} \quad ,
\end{aligned} \tag{2.8}$$

where G has the definition

$$G = 3\ell n[-\frac{1}{3}\kappa^2\phi] - \ell n[\frac{1}{4}\kappa^6|g|^2]. \tag{2.9}$$

Introducing the transformation

$$\phi = -\frac{3}{\kappa^2}\exp(-\frac{\kappa^2}{3}d) \quad , \tag{2.10}$$

Eq. (2.8) reduces to

$$\begin{aligned} L_B = & -(e/2\kappa^2)R(e,\omega) - d_{,a\bar{b}}\, D_\mu Z^{b^+} D^\mu Z^a \\ & + \frac{e}{2}\exp(\frac{\kappa^2}{2} d)[(d^{-1})^{\bar{b}a} G_a G_{\bar{b}} - \frac{3}{2}\kappa^2|g|^2] \\ & + \frac{e}{32} |e_\alpha (d_{,a}(T^\alpha Z)_a)|^2 \\ & - \frac{1}{4}[\mathrm{Tr}(F_{\mu\nu} F^{\mu\nu})] \quad , \end{aligned} \tag{2.11}$$

where

$$G_a \equiv \frac{\partial g}{\partial Z^a} + \frac{\kappa^2}{2} Z^{a^+} g \quad . \tag{2.12}$$

From Eq. (2.11) it is clear that the function $d(Z,Z^+)$ acts as the potential in a Kähler manifold[24] defined by the coordinates Z^a, Z^{b^+} and $(d)_{,ab}$ is the Kähler metric. In the analyses of Ref. [1,21,22] one used

$$d = \frac{1}{2} Z^{a+} Z^a \tag{2.13}$$

which corresponds to the flat Kähler metric $\delta_{a\bar{b}}$ and leads to a normalized kinetic term for the scalar fields in Eq. (2.11). It has been pointed out by Hall, Lykken and Weinberg[9a] that the choice of Eq. (2.13) may be too constraining and perhaps a more realistic form for d is a general function of $Z^{a^+} Z^a$.[25] However, for the analysis of this paper we shall assume Eq. (2.13) and the corresponding tree effective potential[1,21,22]

$$V(Z,Z^+) = \frac{e}{2}\exp(\frac{\kappa^2}{2} Z^{a^+} Z^a)[G_a G_a^+ - \frac{3}{2}\kappa^2|g|^2]$$

$$+ \frac{e}{32} |e_\alpha (Z_a^+, (T^\alpha Z)_a)|^2 \quad . \tag{2.14}$$

III. Gravity Induced Symmetry Breaking

We discuss now the spontaneous breaking in the theory with the tree effective potential defined by Eqs. (2.14). In general the extrema equations arising from Eq. (2.14) have the form

$$(\frac{\partial G_b^+}{\partial Z_a^+} + \frac{\kappa^2}{2} Z_a G_b^+) G_b - \kappa^2 g G_a^+ = 0 \quad . \tag{3.1}$$

If one limits oneself to the real (PC conserving) manifold Eq. (3.1) reduces to[2]

$$T_{ab} G_b = 0, \tag{3.2}$$

where

$$T_{ab} = \frac{\partial^2 g}{\partial Z^a \partial Z^b} + \frac{\kappa^2}{2} (Z_a \frac{\partial g}{\partial Z^b} + Z_b \frac{\partial g}{\partial Z^a})$$
$$+ \frac{\kappa^4}{4} Z_a Z_b \, g - \kappa^2 \delta_{ab} g \quad . \tag{3.3}$$

When G_b vanish in all channels, the solutions to the extrema equations preserve supersymmetry but the gauge group may be broken. When $G_b \neq 0$ at least for one channel,[26] one has a spontaneous breakdown of supersymmetry. To exhibit the phenomena of breakdown of supersymmetry, consider a chiral multiplet $\Sigma = (Z, X_L, h)$ which is a singlet under G and a superpotential $g_2(Z)$ of the form[9]

$$g_2(Z) = m^2 \kappa^{-1} f_2(\kappa Z) \quad , \tag{3.4}$$

where

$$f_2(\kappa^2) = f_2^0 + \kappa Z f_2^1 + . . . \quad . \tag{3.5}$$

In Eq. (3.4) m is a mass parameter and $f_2(\kappa Z)$ is a dimensionless function of the dimensionless quantity κZ which has a smooth expansion in κZ beginning with a constant as in Eq. (3.5). Now if solutions with $G_Z \neq 0$ exist which are minima the satisfaction of Eq. (3.2) requires $T_{ZZ} = 0$. It is then easily seen that nontrivial solutions of Eq. (3.2) imply

$$\langle Z\rangle \sim O(\kappa^{-1}); \ \langle g_2\rangle \sim m^2/\kappa \quad . \tag{3.6}$$

Consequently, if one moves into the unitary gauge and absorbs the spin 1/2 field X into the gravitino field, one obtains a massive gravitino with a mass m_g given by

$$m_g = \frac{\kappa^2}{2} g_2(\langle Z\rangle)\exp(\frac{\kappa^2}{2} \langle Z^+Z\rangle) \sim O(m^2\kappa) \quad . \tag{3.7}$$

We shall identify the gravitino mass shortly as characteristically of the size of the weak interaction mass scale, i.e., $m_g \sim O(m_W)$. Then turning Eq. (3.7) around one has $m \sim (M_W M_{Planck})^{1/2} \sim 10^{10}$GeV. Thus the mass scale appearing in the super-Higgs superpotential of Eq. (3.4) satisfies a geometric hierarchy reminiscent of the analyses in global supersymmetry.[27]

The super-Higgs effect, in particular Eq. (3.7), plays a central role in all discussions of the Supergravity GUT theories.[29] However, one may wonder if the result of Eq. (3.7) is stable when quantum gravity effects are included in our calculations. These quantum loop calculations involving gravitation exchanges and the scalar Higgs particle Z, are obviously very involved, and one needs to define the theory at hand (which is a nonrenormalizable theory in the conventional sense) in an appropriate way such as through the assumption that asymptotic safety[28] holds, to investigate these.

In the following we denote the totality of chiral multiplets of a Supergravity GUT theory by $\Sigma^A = (\Sigma^a, \Sigma^s)$ where Σ^s belong to the "hidden" sector of the Super-Higgs potential (which gives rise to Supergravity breaking) and Σ^a represent the remaining set of GUT fields. The Super-potential for such a theory may in general have the form

$$g(Z^A) = g_1(Z^a) + g_2(Z^s) + g_3(Z^a, Z^s) \quad . \tag{3.8}$$

To protect the GUT sector from acquiring large VEVs of order κ^{-1}, which are characteristic of VEVs in the hidden sector, one generally sets $g_3 = 0$.[31] In the global limit ($\kappa \to 0$) the GUT sector and the hidden sectors are then completely disjoint. However, when one examines the κ dependent effects one finds that each of the sectors is affected by the other. In particular one finds a remarkable phenomenon in that the supergravitational interactions arising from the hidden sector generate soft breaking in the GUT sector. We discuss next a systematic technique for computation of these soft breaking terms.

IV. Gauge Hierarchy and Low Energy Theory

In the investigation of the experimental consequences of the supergravity GUT theory we must determine the corresponding effective theory at low energy. The GUT sector of the theory introduces a new complication in that it can generate effects in the low energy theory not encountered in the couplings of the light fields to supergravity. In SUSY GUT theories, the heavy fields are responsible for the breakdown of the gauge group by acquiring VEV's and masses of order of the GUT mass M (such as the 24 plet $\Sigma^x{}_y$ in SU(5) models[14]). We shall label these fields $Z_I{}^i$. There is, however, another class of heavy fields which also have large masses O(M) but with vanishing VEV's (such as the color triplet Higgs field $H_x{}'$, H^x x = 1, 2, 3 in SU(5) SUSY models[14]). We shall label these fields $Z_{II}{}^i$ and collectively call the heavy fields $Z^i = (Z_I{}^i, Z_{II}{}^i)$. Finally, there are the light fields (we shall label these Z^α) which have zero VEVs and remain massless in SUSY GUT theories with the gauge group broken spontaneously at M. [Masses for the light fields arise in SUSY GUTS only when soft breaking terms are added by hand in the theory.] Thus we have

SUSY GUTS			
Fields	$Z_I{}^i$	$Z_{II}{}^i$	Z^α
VEVs	O(M)	0	0
Masses	O(M)	O(M)	0

(4.1)

Next we consider supergravity GUTS, but to keep things simple, we first assume that the heavy fields Z^i are absent and one has only the fields Z^α and the super-Higgs field Z. Here one finds that Z^α develop a mass of $O(m_g)$ and possibly also a VEV of $O(m_g)$. In realistic models some of these fields would be responsible for the breakdown of electroweak symmetries and thus it is crucial that one protect the mass scale of Z^α when the heavy fields Z^i are included in the analysis. In examining the extrema equations for the light fields (which determine their VEVs) in the presence of the heavy fields, one must eliminate these heavy fields (or integrate them out) from these equations. However, since the heavy fields have the characteristic mass-scale M, there is no a priori guarantee that the effective extrema equations for Z^α would not contain the GUT mass.

It is a remarkable phenomenon that the GUT mass M disappears from the effective tree extrema equations for the light fields Z^α provided the equations

$$\partial^2 g_1/\partial Z^\alpha \partial Z^a \sim O(m_g); \; Z^a = (Z^i, Z^\alpha) \quad , \tag{4.2}$$

hold at the minimum of the potential. Eq. (2.5) puts a severe restriction on the couplings involving $Z_I{}^i$ and Z^α. Thus the coupling $\lambda_{ij} Z_I{}^i Z_I{}^j Z^\alpha$ is permitted only provided

$$\lambda_{ij\alpha} \lesssim O(\frac{m_g}{M}) \quad . \tag{4.3}$$

To show how the ideas outlined above arise, one carries out an expansion in κ of the VEVs Z^a, Z (and similarly of G_B and T_{AB}), so that

$$Z^a = Z^{a(0)} + Z^{a(1)} + \ldots ; \; Z = Z^{(-1)} + Z^{(0)} + \ldots , \tag{4.4}$$

where $Z^{a(n)}$ are the $O(\kappa^n)$ etc. The extrema equations then have the expansion

$$T_{AB}{}^{(0)} G_B{}^{(0)} + [T_{AB}{}^{(0)} G_B{}^{(1)} + T_{AB}{}^{(1)} G_B{}^{(0)}] + \ldots = 0 \quad . \tag{4.5}$$

Using Eqs. (4.2) and (4.5) one can then establish that

$$G_a \sim O(m_g{}^2), \quad G_Z \sim O(m^2) \quad . \tag{4.6}$$

Further, one finds that VEVs and the masses of the various fields satisfy the following expansions:

Supergravity GUTS

The VEV expansions are:

$$Z^i{}_I = 0(M) + 0(m_g) +)(m_g{}^2/M) + \ldots ,$$

$$Z^i{}_{II} = 0(m_g) + 0(m_g{}^2/M) + \ldots ,$$

$$Z^\alpha = 0(m_g) + 0(m_g{}^2/M) + 0(\kappa^2 Mm_g) + \ldots ,$$

$$\kappa Z = 0(1) + 0(m_g{}^2/M) + 0(\kappa^2 Mm_g) + \ldots , \qquad (4.7)$$

[Note, however, that VEVs $Z^i{}_{II} = (H^x, H_x{}')$, $x = 1,2,3$, vanish to all orders as required by color conservation.].

Expansions for the masses are

$$m^2(Z^i) = 0(M^2) + 0(m_g M) + \ldots ,$$

$$m^2(Z^\alpha) = 0(m_g{}^2) + 0(m_g{}^3/M) + \ldots ,$$

$$m^2(Z) = 0(m_g{}^2) + 0(m_g{}^3/M) + \ldots . \qquad (4.8)$$

The protection of the VEVs and masses of the light fields Z^α from the GUT mass M is explicitly seen in Eqs. (4.7) and (4.8). Also from Eqs. (4.7) one finds that the VEVs of the heavy fields Z^i undergo shifts of $0(m_g)$ due to their couplings with the light fields (couplings such as $Z_{II}{}^i Z^\alpha Z^\beta$). Because of these shifts, when the heavy fields are integrated out, one finds a nonvanishing contribution from the heavy GUT sector in the low energy effective theory.

Analysis of the low energy theory can be carried out self-consistently using the expansion of the extrema equations and elimination of the heavy fields as outlined above. It has recently been pointed out by Hall, Lykken and Weinberg,[9a] however, that it is more convenient to deduce first a general form of the low energy effective potential for a GUT theory. The essential elements in the deduction of this potential are the same as discussed in Eqs. (4.2) - (4.8). Here, as in the analysis of the extrema equations, one integrates out the heavy fields. Also the same constraint, i.e.,

Eq. (4.2) at the minimum of the potential V, which guarantee that the GUT mass M does not appear in the extrema equations for the light fields, also guarantees that the low energy effective potential for the light fields does not involve the GUT mass M.

The potential $V = V(Z^i; Z^\alpha; Z)$ satisfies the extrema equations

$$\frac{\partial V}{\partial Z^i} = 0; \quad \frac{\partial V}{\partial Z} = 0 \quad , \tag{4.9}$$

in the heavy and super-Higgs sectors. Using Eqs. (4.9) one may compute $Z^i = Z^i(Z^\alpha)$ and $Z = Z(Z^\alpha)$ and insert these into Eq. (2.14). The resultant low energy effective potential is then[9a]

$$U(Z^\alpha, Z^{\alpha+}) = V[Z^i(Z^\alpha); Z^\alpha; Z(Z^\alpha)] \quad . \tag{4.10}$$

An alternate approach[9b] is to integrate extrema equations Eqs. (3.1) in the light field sectors after heavy fields have been eliminated in these channels using Eqs. (4.9). Since the extrema equations incorporate the protection at low energy from the heavy sector, the effective potential would exhibit the same protection. The explicit form of the effective potential is

$$U(Z^\alpha, Z^{\alpha+}) = \frac{1}{2} e^{\kappa^2/2<Z^+Z>} [\tilde{g}_{1,\alpha}{}^+ \tilde{g}_{1,\alpha} + m_1{}^2 Z_\alpha{}^+ Z_\alpha$$

$$+ \omega + \omega^+ + m_s{}^2(\tilde{g}_{1,i}\bar{G}_i{}^{(0)} + \text{h.c.})] \ , \tag{4.11}$$

where

$$\omega = m_2\tilde{g}_1 + m_3 Z_\alpha g_{1,\alpha} \ , \ m_s \equiv \kappa m^2 \quad , \tag{4.12}$$

$$\tilde{g}_1(Z_i, Z_\alpha) = g_1(Z_i, Z_\alpha) - g_1(Z_i, 0) - b \quad , \tag{4.13}$$

$$m_1{}^2 = \frac{1}{2} m_s{}^2 [\bar{G}_Z{}^{(0)}\bar{G}_Z{}^{(0)*} - \bar{g}_2{}^{(0)}\bar{g}_2{}^{(0)*}] \quad , \tag{4.14}$$

$$m_2 = \frac{1}{2} m_s [\bar{Z}^{(0)}\bar{G}_Z{}^{(0)} - 3\bar{g}_2{}^{(0)}] \quad , \tag{4.15}$$

$$m_3 = \frac{1}{2} m_s \bar{g}_2{}^{(0)} \quad , \tag{4.16}$$

$$Z_i = Z_i^{(0)} - \frac{1}{2} m_s \bar{g}_2^{(0)} (M^{-1})_{ij} A_j^{(0)} \quad , \tag{4.17}$$

and the barred quantities are defined as follows:

$$\kappa Z^{(-1)} = \bar{Z}^{(0)}, \; G_a^{(2)} = m_s^2 \bar{G}_a^{(0)}, \; G_Z = m^2 \bar{G}_Z,$$

$$g_2^{(-1)} = \frac{m^2}{\kappa} \bar{g}_2^{(0)} \quad . \tag{4.18}$$

The definitions of Eq. (4.18) render the barred quantities of O(1). The effective potential of Eq. (4.11) depends on two (possibly complex) mass parameters[32] m_1 and m_2 after one imposes the condition of a zero cosmological term for the theory (which requires $m_1 = |m_3|$).

In the analysis above we have computed the low energy effective potential to leading order and shown that it is protected from the GUT mass, i.e., the potential is $O(m_g^4)$ and terms $O(m_g^3 M)$, etc., are absent. However, it has been pointed out[9b] that in Supergravity GUTS there is a new kind of gauge hierarchy present, due to the appearance of two different mass-scales κ^{-1} and M. Thus, the low energy domain is to be protected not only from the GUT mass M but also from terms of the type $(\kappa M)^n M$, for at least $n \leq 7$. Such terms could arise both at the tree as well as the loop level. It was found in Eqs. (4.7) and (4.8) that terms of this type are all absent at the tree level for models where Eq. (4.2) holds such as for the model considered in Ref. [1]. Eqs. (4.2), however, do not constitute a set of sufficient conditions to guarantee that gauge hierarchy would be maintained as one includes the loop contributions in the effective potential. For a Supergravity GUT theory the form of the one loop effective potential is[3]

$$V^{(1)} = \frac{1}{64\pi^2} \sum_{J=0}^{3/2} (-1)^{2J}(2J+1) \mathrm{Tr} M_J^4 \ln(M_J^2/\mu^2) \quad , \tag{4.19}$$

where M_J is the mass matrix for particles of spin J and μ is the renormalization point. Analysis of the hierarchy, including $V^{(1)}$, shows that other constraints on the superpotential must be satisfied, in addition to those of Eq. (4.2), to maintain the tree hierarchy. The simplest of these new constraints is 3

$$\tilde{g}_{,ab}\tilde{g}_{,ab\alpha} \sim O(m_g) \quad . \tag{4.20}$$

Next we discuss these restrictions, explicitly within the framework of a Supergravity GUT Theory based on G = SU(5) proposed in Ref. [1]. The field content in the g_1 sector of this theory is the same as of the Dimopoulos-Georgi-Sakai model (enhanced by a singlet term) consisting of a $\underline{24}$ $(\Sigma^x{}_y)$, $5(H^x)$, $\bar{5}$ (H_x') and a singlet (U) of Higgs fields and $\bar{5}$ (M_x') and 10 (M^{xy}) of quark-lepton fields. The specific form of g_1 is

$$\begin{aligned} g_1(Z_a) &= \lambda_1(\tfrac{1}{3}\,\mathrm{Tr}\,\Sigma^3 + \tfrac{1}{2}\,M\mathrm{Tr}\,\Sigma^2) \\ &\quad + \lambda_2 H_x'(\Sigma^x{}_y + 3M'\delta^x{}_y)H^y + \lambda_3 U H_x' H^x \\ &\quad + \varepsilon_{\mu\nu\omega xy} H^\mu M^{\nu\mu} f M^{xy} + H_x' M^{xy} g M'_y \end{aligned} \tag{4.21}$$

where f and g are matrices in the generation space. In this scheme, solutions to spontaneous breaking produce three inequivalent vacua, with residual symmetries (i) SU(5), (ii) SU(4) × U(1) and (iii) SU(3) ×SU(2) × U(1). The physically interesting vacua are those belonging to case (iii) and here one adjusts parameters in the superpotential, such that the doublets of Higgs are light. This condition requires setting M′ = M.[33,34] The Higgs fields in the low energy domain then consist of just the singlet field U and the doublets H_α' and H^α (α = 4,5). Including the soft breaking effects due to super-Higgs, one finds then that the electroweak SU(2)×U(1) symmetry breaks down to a residual $U(1)_\gamma$ generating masses for the W and Z mesons proportional to the gravitino mass.

Eq. (4.20) puts severe restrictions on the heavy-heavy-light field couplings. In particular, couplings of the type ΣΣU are forbidden. Also, the couplings $\lambda_3 UH'H$ in Eq. (4.19) would upset the gauge hierarchy [as first noted in Ref. 7] since when a = H´, b = H, and α = U, the right hand side of Eq. (4.20) is O(M) rather than $O(m_g)$. The gauge hierarchy at the one loop level can easily be achieved, however, by the introduction of an additional pair of 5 and $\bar{5}$ (K and K´) of Higgs multiplets where one replaces $\lambda_3 U H_x{}' H^x$ by

$$U(\lambda_3 H_x{}' K^x + \lambda_3{}' K_x{}' H^x) + \lambda_2{}' K'(\Sigma + 3M)K. \tag{4.22}$$

Here Eq. (4.20) as well as the more general condition $V^{(1)}, \alpha \sim O(m_s{}^3)$ are satisfied, guaranteeing one loop stability.

In the models considered above the spontaneous breaking of the electroweak SU(2) × U(1) is achieved at the tree level, but the presence of a singlet field U was required to generate enough structure in the low energy effective potential to allow the electroweak breaking to occur spontaneously. There exists an alternate class of models[4,6,11] where the singlet field is not required to achieve spontaneous breaking of SU(2) × U(1), and the effective potential in the low energy domain depends only on the doublet of Higgs fields H´ and H. The central theme of this approach is to rely on radiative corrections to turn one of the Higgs doublets (the one that couples to the top quark) tachyonic which then triggers spontaneous breaking of SU(2) × U(1). However, since one starts with tree Higgs masses of $O(m_g)$, the radiative corrections needed to achieve negative mass squares must be large, i.e., of the same size as the tree.[36]

V. Low Energy Particle Spectrum

In the analysis of the low energy phenomenology, we shall assume that the Higgs fields in the low energy domain consist of H^α, $H_\alpha{}'$ and U where α = 4,5 as in the analysis of Ref. [1]. In the following discussion we shall assume that SU(2) × U(1) electroweak symmetry breaks at the tree level and hence $\langle H_5{}' \rangle = \langle H^5 \rangle$ in order to minimize the D term in the effective potential. (In scenarios where SU(2) × U(1) breaking is generated through radiative corrections one generally has $\langle H_5' \rangle \ll \langle H^5 \rangle$.[37] We analyze now the spectrum of the various states.

(a) Gaugino and Higgsino Fields

The fermi field content of this theory is as follows: There are eight massless color gluinos which belong to the adjoint representation of SU(3)c, four SU(2) × U(1) gauginos $\lambda^\alpha = \{\lambda^i, i=1,2,3; \lambda^0\}$ and five Higgsinos $\tilde{H}^\alpha$, $\tilde{H}_\alpha'$, $\tilde{U}$ where $\alpha = 4,5$. Thus, in all there are nine low lying fermi fields in the color singlet sector. Diagonalization of the fermi mass matrix shows one massless mode, the photino

$$\lambda^\gamma \equiv \cos\theta_W \lambda^0 + \sin\theta_W \lambda^3 \quad , \tag{5.1}$$

which acts as the supersymmetric partner to the photon (θ_W is the Weinberg angle). Two combinations of the Higgsino fields decouple from the gauginos. The remaining six Higgsino and gaugino fields mix and on diagonalization form two charged Dirac fields $\tilde{W}_{(+)}$ and $\tilde{W}_{(-)}$ with masses $\tilde{m}_{(+)}$ and $\tilde{m}_{(-)}$ lying above and below the $W^\pm$ meson mass, and two Majorana fields $\tilde{Z}_{(\pm)}$ with masses $\tilde{\mu}_{(\pm)}$ lying above and below the Z^0 mass. The Wino fields are the following linear combinations of the Higgsino and gaugino fields:

$$\tilde{W}_{(\pm)} = f_\pm X_+ \mp f_\mp \lambda_+ ; \; f_\pm \equiv (\tilde{m}_{(\pm)}/(\tilde{m}_{(+)} + \tilde{m}_{(-)}))^{1/2} \quad , \tag{5.2}$$

where $\sqrt{2}\,\lambda_+ = (\lambda^1 + i\lambda^2)$ and $X_+ = i(H_4' + H^{4C})$. The Wino masses are given by[13b, 38]

$$\tilde{m}_{(\pm)} = \sqrt{M_W^2 + m_1^2} \pm m_1 \quad , \tag{5.3}$$

where M_W is the W mass and m_1 is a model dependent parameter.[39] A similar analysis holds for the Zino masses and one has

$$\mu_{(\pm)} = \sqrt{M_Z^2 + m_1^2} \pm m_1 \quad , \tag{5.4}$$

where M_Z is the Z meson mass. Eqs. (5.3) and (5.4) then yield the following mass relations :

$$\tilde{m}_{(+)} - \tilde{m}_{(-)} = \tilde{\mu}_{(+)} - \tilde{\mu}_{(-)} \quad , \tag{5.5}$$

$$\tilde{m}_{(+)}\tilde{m}_{(-)} = M_W^2 \; ; \; \tilde{\mu}_{(+)}\tilde{\mu}_{(-)} = M_Z^2 \quad . \tag{5.6}$$

(b) Higgs and Vector Mesons

There are five complex or ten hermitian scalar fields in the low energy theory from the multiplets H^α, $H_\alpha{}'$ and U and four SU(2) × U(1) gauge boson, $V_\mu{}^\alpha = \{V_\mu{}^i, i=1,2,3; V_\mu{}^o\}$. Upon diagonalization one finds one massless spin 1 (the photon) and three massless spin zero Goldstone particles which are absorbed by $W^\pm$ and Z^0 to grow masses. The remaining 7 spin zero mesons consist of one charged state with mass $m_H{}^+$ larger than M_W, one neutral state with mass $m_H{}^0$ larger than M_Z obeying the relation[9a,13b,40]

$$m_H{}^{+2} - M_W{}^2 = M_H{}^{02} - M_Z{}^2 = \bar{m}_H{}^2 \quad , \tag{5.7}$$

and four additional neutral states which involve components of H^5, $H_5{}'$, and U. One may also derive a sum rule relating the Higgs, W meson, Wino and gravitino masses. One has

$$\frac{1}{2} m_H{}^{+2} = m_g{}^2 + \frac{1}{2} M_W{}^2 + (M_W{}^2 - \tilde{m}_{(-)}{}^2)^2/\tilde{m}_{(-)}{}^2 \quad . \tag{5.8}$$

(c) Sleptons and Squarks

In the matter sector quarks and leptons would grow masses as usual due to their Yukawa couplings. Now each massive quark (q_i) and lepton (ℓ_i) has two complex scalar superpartners which we label $\tilde{q}_\pm$ and $\tilde{\ell}_\pm$ arising because each Weyl spinor has one complex scalar superpartner. (Hence each massless neutrino has a single scalar partner, $\tilde{\nu}$.) The mass spectrum for the Bosonic superpartners is as follows[9a, 13b]:

$$m(\tilde{q}_{\pm})^2 = m_{\tilde{\nu}}^{\ 2} + m_q^{\ 2} \pm \beta m_{\tilde{\nu}} m_q \quad , \tag{5.9}$$

$$m(\tilde{e}_{\pm}) = m(\tilde{d}_{\pm}); \quad m_{\tilde{\nu}} = m_g \quad , \tag{5.10}$$

where β in Eq. (5.9) is a model dependent parameter.[41] We note that the superpartners of the neutrinos have the same universal mass, the mass of the gravitino, while the masses of the superpartners of the massive quarks and leptons have splittings from the universal mass proportional to the corresponding quark and lepton masses. The formulae Eqs. (5.9) and (5.10) hold at the GUT mass and there would be renormalization effects at low energy on the individual masses. Since the quark masses are generally small compared to m_g, one finds an approximate degeneracy in the masses of their scalar superpartners which also leads to a suppression of the flavor changing neutral currents.

(d) Loop Masses of Photino and Gluinos

As noted already the photino and gluinos are massless at the tree level. However, at the loop level the exchange of heavy fields, i.e., the <u>24</u> plet of scalars $\Sigma^x{}_y$ and their spin 1/2 superpartners generate non-vanishing contributions to the masses for the entire <u>24</u> plet of gauge bosons. The relevant vertices for the loop are generated by the supersymmetric counterparts of the Yang-Mills interactions

$$L_{\lambda XZ} = + \frac{i}{2} g Z T_\alpha \bar{X} \lambda^\alpha + \text{h.c.} \quad , \tag{5.11}$$

where Z represent the heavy scalars and X their superpartners, T_α are the generators and λ^α, the gauginos. The masses for the photino and the gluinos are[42,11]

$$m_{\tilde{\gamma}} = \frac{8}{3}\left(\frac{\alpha}{4\pi}\right)\bar{C}m_g; \quad m_{\tilde{g}} = \frac{\alpha_s}{4\pi}\bar{C}m_g , \tag{5.12}$$

where $\bar{C}$ is proportional to the Casimir of the heavy field being exchanged, α is the e.m. and α_s is the QCD coupling constant. The photino mass is expected to lie in the range of (1-5) GeV and the gluino mass in the range (5-25) GeV from Eq. (5.12). The $\tilde{W}_\pm$ Winos and $\tilde{Z}_\pm$ Zinos would also receive contributions from the loop terms in addition to their tree masses.

VI. Experimental consequences of Supersymmetry and of Supergravity GUTS

Supersymmetry predicts an array of new odd R parity objects, such as the photino, gluinos, selectrons, squarks, Higgsinos, Winos, and Zinos. In Supergravity GUTS the masses of these particles are determined in terms of the mass scale fixed by the gravitino mass which is of the order of the weak interaction mass scale, i.e., $O(M_W)$. We note that the entire mass spectrum is supergravitationally induced. Thus, in part the masses arise from the super-Higgs effect and in part from the gravitationally induced SU(2) × U(1) breaking. The mass spectrum generated this way leads to some interesting decay patterns for some of these particles. We discuss the phenomenology of some of these which are most propitious for observation in the current machines PEP. PETRA, and SPS, and in the proposed machines at LEP, SLC AND HERA.

Photinos and Selectrons

In the Supergravity GUTS photino is expected to be the lowest mass odd R parity object and hence is expected to be absolutely stable. In e^+e^- collisions one may produce the photinos in pairs[43]

$$e^+e^- \rightarrow \tilde{\gamma}\tilde{\gamma}, \tilde{\gamma}\tilde{\gamma}\gamma, \ldots \quad . \tag{6.1}$$

If the photino is indeed the lowest mass odd R parity object, the $\tilde{\gamma}\tilde{\gamma}$ final state would be unobservable and the relevant process experimentally is the $\tilde{\gamma}\tilde{\gamma}\tilde{\gamma}$ final state where one would observe a single soft photon with missing mass. The background for this process is $e^+e^- \to \nu\bar{\nu}\gamma$ but the background is expected to decrease at higher energies.[43] These experiments can be carried out with current energies at PEP and PETRA. Production of selectrons would require higher energies like those contemplated at LEP and HERA and the production would proceed through[44]

$$e^+e^- \to \tilde{e}^- + \tilde{\gamma} + e^+ \quad , \tag{6.2}$$

$$e^-p \to \tilde{e}^- + \tilde{\gamma} + X \quad . \tag{6.3}$$

The current lower limit on selectron mass from PETRA is $\gtrsim$ 20 GeV.[45]

Winos and Zinos

As discussed in Section V, perhaps the most dramatic result of supersymmetry and of Supergravity GUT theory is the prediction of the existence of a Wino and a Zino below the W and Z meson masses. We discuss now the decays of the W and Z mesons into their supersymmetric partners as a consequence of this phenomena. The W can decay into a Wino and a photino and into a Wino and a Zino unless these processes are forbidden by kinematics. The branching rations for these decays relative to $W \to e\nu$ are

(i) If $M_W > \tilde{m}_{(-)} + m_{\tilde{\gamma}}$ then one has[13a] ,

$$\Gamma(W \to \tilde{W}\tilde{\gamma})/\Gamma(W \to e\nu) = 4\,\mathrm{Sin}^2\theta_W f_-\Phi_1 \quad ;$$

(ii) If $M_W > \tilde{m}_{(-)} + \tilde{\mu}_{(-)}$[46] ,

$$\Gamma(W \to \tilde{W}\tilde{Z})/\Gamma(W \to e\nu) = (4\,\mathrm{Cos}\theta_W f_+ g_+ + f_- g_-)^2 \Phi_2 \quad ;$$

where $f_\pm$ are defined by Eq. (5.2) and $g_\pm$ are defined analogous to

$f_{\pm}$ with $\tilde{m}_{(\pm)}$ replaced by $\tilde{\mu}_{(\pm)}$, Φ_1 and Φ_2 are phase space factors and θ_W is the Weinberg angle. For a Wino mass $\tilde{m}_{(-)} = 30$ GeV and hence a corresponding Zino mass $\tilde{\mu}_{(-)} = 37$ GeV one has

$$\Gamma(W \to \tilde{W}\tilde{\gamma})/\Gamma(W \to e\nu) \simeq 0.6 \tag{6.4}$$

and

$$\Gamma(W \to \tilde{W}\tilde{Z})/\Gamma(W \to e\nu) \simeq 2.2 \quad . \tag{6.5}$$

Next we consider the Z meson which can decay into two Winos unless forbidden by kinematics. The branching ration $Z \to \tilde{W}^+\tilde{W}^-$ relative to $Z^0 \to e^+e^-$ is [13b,47]

(iii) $M_Z > 2\tilde{m}_{(-)}$

$$\Gamma(Z^0 \to \tilde{W}^+\tilde{W}^-)/\Gamma(Z^0 \to e^+e^-) = 16(\mathrm{Cos}\theta_W)^4[1 - \frac{1}{2}(\frac{f^-}{\mathrm{Cos}\theta_W})^2]^2\Phi_2$$

$$\sim 7.6 \quad , \tag{6.6}$$

where Φ_2 is a phase-space factor and the branching ratio is computed for the same parameters as Eqs. (6.4) and (6.5). We note that the branching ratio of Eq. (6.6) is enormous and thus if kinematically allowed, the $Z^0 \to \tilde{W}^+\tilde{W}^-$ process would be a prominent decay mode of the Z^0.

The Winos and Zinos produced in the decays of the W and Z are of course unstable. Thus, a Wino can have a sequential decay through a W or through a squark (slepton) intermediate state:

$$\tilde{W} \to \tilde{\gamma} + W \to \tilde{\gamma}\ell\nu_\ell,\ \tilde{\gamma}q_i\bar{q}_j \tag{6.7}$$

and

$$\tilde{W} \to \tilde{q}_i + \bar{q}_j \to \tilde{g}\, q_i \bar{q}_j, \text{ etc.} \quad . \qquad (6.8)$$

In Eqs. (6.7) and (6.8) the final state quarks and gluinos would jet and hadronize. In a similar way the Zino can have a sequential decay through a squark (slepton) or a W intermediate state with final states as follows:

$$\tilde{Z} \to \tilde{W} + W \to \tilde{W}q\bar{q},\ \tilde{W}\ell\nu_\ell \text{ etc.} \quad , \qquad (6.9)$$

$$\tilde{Z} \to \tilde{q} + \bar{q} \to q\bar{q}\ \tilde{g} \quad , \qquad (6.10)$$

where in Eq. (6.9) the final Wino state would have further sequential decays according to Eqs. (6.7) and (6.8). Which of these decay channels would dominate depends on the squark masses versus the W meson mass ratios. Thus, if the squark masses are large compared to the W meson mass (i.e., $m_{\tilde{q}} \gtrsim 3\, M_W$), the channels with W meson intermediate state dominate i.e., decays of Eq. (6.7) dominate over decays of Eq, (6.8) and decays of Eq. (6.9) dominate over decays of Eq. (6.10). However, if the squark masses are comparable to the W meson mass, then the decays of Eq. (6.7) are comparable to decays of Eq. (6.8) and decays of Eq. (6.9) are comparable to decays of Eq. (6.10).

Using Eqs. (6.7) - (6.10), we look for signals in W and Z decays which would indicate these decays as arising through the Wino or Zino intermediate states. A characteristic of these decays are multiparticle final states with many (unobservable) neutrals. Some of the interesting decay signals are shown below with preliminary calculations of the branching ratio. The $W^\pm$ branching ratios are relative to the $W^+ \to e^+ + \nu_e$ while the Z^0 branching ratios are relative to $Z^0 \to e^+e^-$. The calculations are for the same parameters as used in Eqs. (6.4) - (6.6).

Final State	Relative Branching Ration %
(A) $W \to \tilde{\gamma} + \tilde{W}$ Mode	
$[\tilde{\gamma}] + (2\text{jets} + [\tilde{\gamma}])$	25
$[\tilde{\gamma}] + (3\text{jets} + [\tilde{\gamma}])$	25
(B) $W \to \tilde{W} + \tilde{Z}$ Mode	
$(\ell + [\nu_\ell + \tilde{\gamma}]) + (\text{hadrons})$	30
$(\text{hadrons} + [\tilde{\gamma}])$	
$+ (\ell_1 + \ell_2 + [\nu_1 + \nu_2 + \tilde{\gamma}])$	$\lesssim 1$
$\ell_1 + \ell_2 + \ell_3 + [\nu_1 + \nu_2 + \nu_3 + 2\tilde{\gamma}]$	< 1
(C) $Z \to \tilde{W}^+ + \tilde{W}^-$ Mode	
$(\ell + [\nu + \tilde{\gamma}]) + (3\text{jets} + [\tilde{\gamma}])$	95
$(\ell + [\nu + \tilde{\gamma}]) + (2\text{jets} + [\tilde{\gamma}])$	95

The particles in the square bracket in each of the decays above are the unobserved neutrals. In the mode (A) decays the jets that come out in the final state would exhibit large unbalanced momentum, while the single charged final state lepton of mode (B) and the hadrons would be in opposite hemispheres. The most distinctive decay mode is the three charged lepton final state of mode (B), though it has a rather low branching ratio. The supersymmetric decay modes of Z^0 are rather large and should be obervable in abundance at LEP or SLC. Again, the leptons should be mainly in opposite hemispheres to the jets.

Acknowledgements

We are pleased to acknowledge conversations with C.S. Aulakh, L. Hall, J. Polchinsky, D. Shambroom, E. VonGoeler and S. Weinberg.

References

1. A.H. Chamseddine, R. Arnowitt and P. Nath, Phys. Rev. Lett. 49, 970 (1982).
2. P. Nath, R. Arnowitt and A.H. Chamseddine, Phys. Letters 121B, 33 (1983).
3. R. Arnowitt, A.H. Chamsiddine and P. Nath, Phys. Letters 120B, 145 (1983).
4. L.E. Ibanes, TH, 3374-CERN (1982); Universidad Autonoma de Madrid preprint, FTUAMI 82- 8 (1982).
5. R. Barbieri, S. Ferrara and C.A. Savoy, Phys. Lett. 119B, 353 (1982); H.P. Nilles, M. Srednicki and D. Wyler, TH 3432-CERN (1982).
6. J. Ellis, D.V. Nanopoulos and K. Tamvakis, Phys. Lett. 121B, 1231 (1983); J. Ellis, J.S. Hagelin, D.V. Nanopoulos and K. Tamvakis, SLAC-POB-3042 (1983).
7. H.P. Nilles, M. Srednicki and D. Wyler, TH. 3461-CERN (1982); B. Lahanas, TH. 3467-CERN (1982).
8. S. Ferrara, D.V. Nanopoulos and C.A. Savoy, TH. 3442-CERN (1982).

9a. L. Hall, J. Lykken and S. Weinberg, University of Texas Report No. UTTG-1-83.

9b. P. Nath, R. Arnowitt and A.H. Chamseddine, NUB#2579 (1982)/ HUTP-82/A057.

10. N. Ohta, Tokyo-Preprint UT-388; C.S. Aulakh, CCNY-HEP-83/2; J. Leon, M. Quiros and M. Ramon Medrano, Madrid Preprint.
11. L. Alvarez-Gaume, J. Polchinski and M.B. Wise, HUTP-821A063/ CALT-68-990.
12. E. Cremmer, P. Fayet and L. Girandello, University of Paris preprint LPTENS-82/30 (1982); S.K. Soni and H.A. Weldon, University of Pennsylvania preprint (1983).

13a. S. Weinberg, Phys. Rev. Letters 50, 387 (1983).

13b. R. Arnowitt, A.H. Chamseddine and P. Nath, Phys. Rev. Letters 50, 232 (1983).

14. E. Witten, Nucl. Phys. B177 (1981); B185, 513 (1981); M. Dine, W. Fishler and M. Srednicki, Phys. Lett. 104B, 199 (1981); S. Dimopoulos and H. Georgi, Nucl. Phys. B193, 150 (1981); N. Sakai, Z. für Phys. C11, 153 (1981); D.V. Nanopoulos and K. Tamvakis, CERN preprints TH. 3327, 3247 (1982); R.K. Kaul, Phys. Lett. 109B, 19 (1982); C.S. Aulakh and R.H. Mohapatra Preprint CCNY-HEP-82/4; C. Nappi and V. Ovrut, Phys. Lett. 113B, 175 (1982); L. Alvarez-Gaume, M. Claudson and M.B. Wise, Nucl. Phys. B207, 96 (1982).

15. For a recent review of globally sypersymmetric theories see P. Fayet, in proceedings of the 21st International Conference on High Energy Physics, Paris 26-31 July 1982, Journal de Physique C3-673.

16. J. Wess and B. Zumino, Nucl. Phys. B70, 39 (1974).

17. D.Z. Freedman, P. van Nieuwenhuizen and S. Ferrara, Phys. Rev. 1013, 3214 (1976); S. Deser and B. Zumino, Phys. Lett. 62B, 335 (1976).

18. S. Ferrara and P. van Nieuwenhuizen, Phys. Lett. 76B; J. Stelle and P. West, Phys. Lett. 77B, 376 (1978); 74B, 330 (1978); Nucl. Phys. B145, 175 (1978).

19. We use conventions of P. van Nieuwenhuizen, Phys. Reports 68(4), 189-398 (1981).

20. E. Cremmer, B. Julia, J. Scherk, S. Ferrara, L. Girardello and P. van Nieuwenhuizen, Nucl. Phys. B147, 105 (1979)

21. E. Cremmer, S. Ferrara, L. Girardello and A. Van Proeyen, Phys. Lett. 116, B231 (1982) Th. 3348-CERN (1982).

22. J. Bagger and E. Witten, Phys. Lett. 118B, 103 (1982); J. Bagger, Nucl. Phys. B211, 302 (1983).

23. $(d^{-1})^{ba}$ appearing in Eq. (2.11) is the inverse of the matrix $d_{,a\bar{b}}$.

24. B. Zumino, Phys. Lett. 87B, 203 (1979).

25. Such a choice is motivated in part by the reasoning that supergravity loop corrections would maintain an approximate U(n) symmetry among its n chiral superfields if the matter couplings are small.[9a] Thus, one may simulate gravitational quantum loop corrections in scalar potential by choosing a general function of $Z^a Z^{a+}$ for $d(Z,Z^+)$.

26. In this case, T_{ab} has at least one zero eigenvalue.

27. S. Dimopoulos and S. Raby, Los Alamos Preprint LA-UR-82-1282; J. Ellis, L. Ibanes and G. Ross, Rutherford Preprint RL 82-024 (1982); J. Polchinsky and L. Susskind, Phys. Rev. D26, 3361 (1982).

28. S. Weinberg, in General Relativity--An Einstein Centenary Survey, edited by S.W. Hawking and W. Israel (Cambridge University Press, Cambridge, England, 1979), Chapter 16.

29. The simplest super-Higgs potential is a linear one defined by[30] $g_2(Z) = m^2(Z+B)$. The minima correspond to $\kappa Z_{(-1)} = a(\sqrt{2} - \sqrt{6})$; $\kappa B_{(-1)} = -a(2\sqrt{2} - \sqrt{6})$; $a = \pm 1$.

30. J. Polony, University of Budapest Report No. KFKI-1977-93, 1977 (unpublished); See also Ref. 20.

31. Actually one may maintain the desired protection and allow a mixed term in the superpotential e.g.,

$$g_3(Z^a,Z^s) = \lambda^1{}_{abs}\, Z^a Z^b Z^s + \lambda^2{}_{ars}\, Z^a Z^r Z^s$$

if the couplings $\lambda^1{}_{abs}$ are $O(\kappa m_g)$ and $\lambda^2{}_{ars}$ are $O(\kappa^2 m_g^2)$.

32. There are four mass parameters in the low energy effective potential for the case where one has a general Kähler metric in the original potential.[9a]

33. The necessity of fine tuning to achieve light Higgs doublets cannot be circumvented in an obvious way in this model. We thank C.S. Aulakh and J. Polchinsky for discussions on this question.

34. It is straightforward to construct global models where the Higgs doublets are naturally light using the "missing partner" mechanism. See in this context, B. Grinstein, Nucl. Phys. B206, 387-396 (1982); A. Masiero, D.V. Nanopoulos, K. Tamvakis and T. Yanagida, Phys. Lett. 115B, 380 (1982); H. Georgi, Phys. Lett. 108B, 283 (1982); S. Dimopoulos and H. Georgi, HUTP-82/A046 (1982).

35. A similar suggestion has also been made by Ferrara et al. in Ref. 8 in the context of the missing partner scenario.[34]

36. For this purpose one needs a large mass of the top quark of order $\gtrsim$ 100 GeV.

37. A more general phenomenological analysis which can interpolate between the formalism where spontaneous breaking occurs at the tree level and the one where spontaneous symmetry breaking is triggered through radiative corrections can be given, and shall be presented elsewhere.

38. A model independent prediction of the existence of a Wino below the W meson mass and a Zino below the Z meson mass is given by S. Weinberg.[13a]

39. For the case of the linear super-Higgs model[29] m_1 has the value $m_1 = \frac{1}{2} m_g(3\lambda - x)$ where $\lambda = \lambda_2/\lambda_1$ and $U \equiv - m_s x/(\sqrt{2}\ \lambda_3)$.

40. $\bar{m}_H$ is a model dependent quantity proportional to the gravitino mass and for the linear super-Higgs model it has the value

$$\bar{m}_H = m_g \sqrt{2}[1 + (3\lambda - x)^2]^{1/2}$$

as discussed in Ref. (13b).

41. For the model of Ref. 29, β has the value $\beta = \sqrt{2}(-x + 3\lambda - 3 + \sqrt{3})$. In the more general analysis of Ref. 9a, allowing also for a curved Kähler manifold $m_{\tilde{\nu}}$ need not have the value m_g.

42. R. Arnowitt, C.H. Chamseddine and P. Nath (unpublished).

43. P. Fayet, Ecole Normale Superiere preprint LPTENS-82/12 (1982); J. Ellis and J.S. Hagelin, SLAC-PUB-3014 (1982).

44. P. Salati and J.C. Wallet, LAPP-TH-65 (1982).

45. R. Brandelik et al., Phys. Lett. 117B, 365 (1982).

46. P. Nath, R. Arnowitt and A.H. Chamseddine, "Wino and Zino Decays of the W and Z Mesons," NUB#2588.

47. There is an additional factor of 4 omitted in Eq. (16) of Ref. 13b and Eq. (47) of Ref. 48.

48. P. Nath, R. Arnowitt and A.H. Chamseddine, "Global and Local Supersymmetric Grand Unification," NUB#2586.

DYNAMICAL SYMMETRY BREAKING: A STATUS REPORT*

M.A.B. Bég*

The Rockefeller University

New York, New York 10021

Abstract

An overview of the problems afflicting the subject, of dynamical symmetry breaking predicated on the concept of hypercolor, is followed by a description of the $\Delta S=2$ conundrum and a brief exegis of some recent work. The report concludes on a cautiously optimistic note.

1. State of Dynamical Symmetry Breaking

The subject of dynamical symmetry breaking (DSB) in weak interaction theory, based on the concept of hypercolor,[1] has been in a state of suspended animation for some time. Factors contributing to inhibition of activity include:

(i) A surge of interest in supersymmetry[2] (SUSY).

By providing a theoretical framework in which there is a a natural place for spin-0 fields, SUSY deprives hypercolor of its prime raison d'être.

(ii) Intractability of associated mathematical problems.

Despite the paucity of parameters, which characterizes a

*Work supported in part by the Department of Energy under Contract Grant No. DE-AC02-83ER40033.B000.

a dynamical theory, the fact remains that we are able to calculate almost nothing. This has to do with our inability, to date, to unlock the complex mathematical structure of QCD-like theories in the infrared sector. This sector, with its essential singularities in the coupling constant plane -- which preclude the use of any perturbative techniques -- is where the physics of hypercolor-based DSB resides.

(iii) The problem of flavor changing neutral currents (FCNC's).

Extensions of the hypercolor scenario, proposed to date, which permit leptons and quarks to have the attribute of mass (for quarks I mean masses which simulate current masses in the low energy limit) do not exhibit the GIM mechanism of the canonical theory. As a result, unacceptably large flavor changing amplitudes -- $\Delta S=2$ processes are a particularly acute source of embarrassment -- can result if the parameters of the theory are adjusted so that the masses of the heavy quarks (s-quark or heavier) come out correctly. This problem, first noticed by Dimopoulos and Ellis,[3] has been recently cited by Fayet[4] and Salam[5] as a serious difficulty for the dynamical construct.

(iv) Intimations of a direct clash with experiment.

Many authors have suggested that low-lying pseudo-Goldstone bosons (PGB's) may be a characteristic feature of the hypercolor scenario.[6] Experimental searches[7] for charged hyperpions, by two groups working at the PETRA storage ring, have, however, failed to reveal any such objects. The TASSO group, investigating hadronic decay modes, concluded at 95% C.L. that there are no PGB's in the mass range 5 GeV to 13 GeV; this has been cited as evidence against hypercolor.[8] It is difficult to draw any firm conclusions, however, because PGB masses do not lend themselves to precise computation. Indeed, it has already been noted in the literature[9] that the above mass values may be too low, that a meaningful quest for PGB's requires scanning the mass range up to ~ 100 GeV. Very little can be said until machines such as LEP go into operation.[10]

(v) Elusiveness of a working model.

No model which reproduces physical reality is yet in hand; the situation is in marked contrast to that in the canonical theory.

I have nothing further to say about (i), (ii), (iv), or (v), but I shall comment briefly on (iii).

2. The FCNC Conundrum

The seeds of the conundrum lie in the mechanism, proposed by several authors,[11] for quark mass generation, wherein one feeds the dynamical hyperquark mass into the quark sector with the help of subelectroweak interactions characterized by a mass scale of the order of 10 TeV. [See Fig. 1, reproduced from my report to Orbis Scientiae '80].

The interactions operative in mass generation are deemed to emerge from a gauge group, preferably one that commutes with the electroweak group; to have nonvanishing values for s and d masses, the group must contain the generators:

$$Q(s) = \int d^3x \;\; s_w^{\dagger} q' \quad , \tag{1}$$

$$Q(d) = \int d^3x \;\; d_w^{\dagger} q' \quad , \tag{2}$$

where q' is a hyperquark and the w-suffix indicates that we are

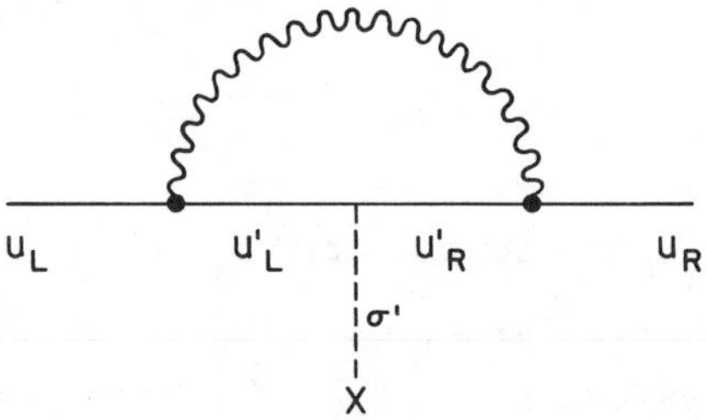

Fig. 1. Possible mechanism for generating current mass for the u-quark. The wavy line represents one of the heavy bosons which mediate subelectroweak interactions.

working with weak -- as opposed to mass -- eigenfields. Closure of the Lie algebra then requires that it also contain the generators:

$$Q_1 = [Q(s), Q(d)^\dagger] = \int d^3x \ s_w^\dagger d_w \quad , \tag{3}$$

$$Q_2 = Q_1^\dagger = \int d^3x \ d_w^\dagger s_w \quad , \tag{4}$$

$$Q_3 = [1/2] \int d^3x \ (s_w^\dagger s_w - d_w^\dagger d_w) \quad . \tag{5}$$

These generators imply FCNC's and consequent $\Delta S=2$; indeed, they account for the worst of the $\Delta S=2$ problem through the single gauge boson exchange process, dubbed SGEX in Ref. 3. Making many assumptions, which need not be reviewed here, the authors of Ref. 3 found:

$$L_{\text{eff.}}^{\Delta S=2} \text{(Hypercolor)} \approx 10^3 \ L_{\text{eff.}}^{\Delta S=2} \text{(Canonical)} + \text{(RR and LR)} \tag{6}$$

for

$$m_s \approx 200 \text{ MeV} \quad . \tag{7}$$

Here (RR and LR) indicates contributions to $L_{\text{eff.}}$ which are absent in the canonical theory; they stem from the coupling of right-handed currents to themselves and to left-handed currents; as is evident, both currents are required for mass generation in the dynamical construct.

The $\Delta S=2$ problem has been recently reexamined[12] in the context of a model suggested by the horizontal group[13] G_H, which commutes with the electroweak group:

$$G_H = SU(n)_L^{\text{u and d-quarks}} \otimes SU(n)_R^{\text{u-quarks}} \otimes SU(n)_R^{\text{d-quarks}}$$

$$\otimes\; SU(n)_L^{\text{u and d-leptons}} \otimes SU(n)_R^{\text{d-leptons}} \otimes U(1). \tag{8}$$

Note that the "n" in the leptonic groups will be identical to the "n" in the quark groups only if color-like degrees of freedom remain dormant and are not counted.

We are concerned only with current quark masses, however, and may therefore gauge the group:

$$G_H^{\text{quarks}} = SU(n)_L^{\text{u and d}} \otimes SU(n)_R^{u} \otimes SU(n)_R^{d} \tag{9}$$

with

$$n = 6 + n_F' \, n_C' \tag{10}$$

where n_C' is the number of hypercolors and n_F' is the number of u-type or d-type hyperflavors. The results stated below emerge from analysis based on the following assignments of quarks and hyperquarks to the fundamental representations of the groups in Eq. (9):

$$SU(n)_L^{\text{u and d}} : \quad \begin{pmatrix} d_{wa} \\ s_{wa} \\ d_b' \end{pmatrix}_L \oplus \begin{pmatrix} u_{wa} \\ c_{wa} \\ u_b' \end{pmatrix}_L , \tag{11}$$

$$SU(n)^{d}_{R} : \begin{pmatrix} d_{wa} \\ s_{wa} \\ d'_{b} \end{pmatrix}_R \quad ; \quad SU(n)^{u}_{R} : \begin{pmatrix} u_{wa} \\ c_{wa} \\ u'_{b} \end{pmatrix}_R \quad . \tag{12}$$

Here a is the color index (a = 1,2 or 3) and b labels both hypercolor and hyperflavor ($b = 1,2, \ldots, n'_C\, n'_F$). We have simplified the discussion by considering only four flavors of ordinary quarks, and by requiring that hyperquarks do not have the attribute of ordinary color. The weak eigenfields are related to mass eigenfields in the usual way:

$$\begin{pmatrix} d_w \\ s_w \end{pmatrix} = \begin{pmatrix} \cos\theta_1 & \sin\theta_1 \\ -\sin\theta_1 & \cos\theta_1 \end{pmatrix} \begin{pmatrix} d \\ s \end{pmatrix} , \tag{13}$$

$$\begin{pmatrix} u_w \\ c_w \end{pmatrix} = \begin{pmatrix} \cos\theta_2 & \sin\theta_2 \\ -\sin\theta_2 & \cos\theta_2 \end{pmatrix} \begin{pmatrix} u \\ c \end{pmatrix} , \tag{14}$$

with

$$\theta = \theta_1 - \theta_2 \tag{15}$$

being Cabibbo's angle, the angle relevant to low-energy physics.

Now almost any scenario for breaking G_H^{quarks} down to $SU(3)_C \otimes SU(n'_C)_{C'}$ leads to a vast assortment of mixing angles in the gauge field sector. Some of these are directly constrained by experiment, others are not. Many of these unconstrained angles are, of course, calculable in principle; present technology, however, is not adequate for the task. Until better nonperturbative calculational techniques become available, the sensible thing to do is to carry these angles as free parameters. The following is within the conceptual framework afforded by this standpoint.

Results mentioned earlier may now be stated:

(a) The SGEX contribution, to $\Delta S=2$, depends on unconstrained angles and thus, at this time, may be deemed to be manually controllable.

(b) The SGEX part is also subject to a measure of natural suppression.

(c) Higher order exchange (HOEX) contributions to $\Delta S=2$, specifically the parts stemming from double gauge boson exchange, are safely small.

Discussion of (c) is necessarily somewhat technical and I must refer you to a forthcoming paper.[14] To see (a) and (b), consider a part of the Lagrangian:

$$\delta L = g_2[\bar{D}_w(\theta_1)\gamma_\mu(1/2)\tau_i D_w(\theta_1)]_R \cdot E_R^{\mu,i} \tag{16}$$

where the $E_R^{\mu,i}$ are color singlet gauge fields and

$$D_w(\theta_1) = \begin{pmatrix} d_w \\ s_w \end{pmatrix} . \tag{17}$$

Now $E_R^{\mu,1}$ and $E_R^{\mu,3}$ can mix, and if this mixing angle be 2χ, we find that the quarks which couple to gauge fields of definite mass are $D_w(\theta_1-\chi)$.

The effective $\Delta S=2$ Lagrangian, which arises from Eq. (16), is then:

$$\delta L_{\text{eff.}}^{S=2} = (g_2^2/4M^2) \quad (\delta M^2/M^2) \quad \sin^2 \; 2(\theta_1-\chi)$$

$$(\bar{d}_R\gamma_\mu s_R) \cdot (\bar{d}_R\gamma^\mu s_R) \; + \; \text{H.C.} \quad , \tag{18}$$

where $\delta M^2 = M(\bar{E}^{\mp})^2 - M(\bar{E}^3)^2$ and $M^4 = M(\bar{E}^{\mp})^2 \cdot M(\bar{E}^3)^2$, $E^{\mp}$ $(\equiv \bar{E}^{(1\mp i\ 2)})$ and $\bar{E}^3$ being the mass eigengields.

Natural suppression -- the phenomenon dubbed pseudo-GIM in

Ref. 11 -- arises from the $\delta M^2/M^2$ factor in Eq. (18). [The analogous ratio in the Weinberg-Salam theory would be $(M_z^2 - M_w^2)/M_zM_w \simeq 0.20$]. While δM^2 may be quite small compared to M^2, the actual limit $\delta M^2=0$ corresponds to the unphysical situation in which there is an SU(2) symmetry such that $m_s = m_d$, $m_c = m_u$. Note that the angle χ is a prime example of an unconstrained, adjustable, angle.

Let me conclude this part of the discussion by emphasizing that the work described above can not, by any stretch of the imagination, be regarded as complete. We have not, for example, addressed ourselves to the question of mass generation for leptons. [See, however, Ref. 11, for a proposal to resolve the lepton mass problem.]

3. Outlook

If we are obliged to abandon hypercolor, it would have to be because of the intractability of the mathematical problems posed by the scheme. or by appeal to aesthetic criteria, not because of any clash with low energy experiments. And if PGB's are discovered at LEP, or if something like a hyperhadron spectrum is seen to open up at the Tevatron or the next generation of hadron colliders, it might vindicate the underlying ideas and force us to view the problems afflicting the subject in a very different light.

References

1. Reviews which afford perspectives on the subject and contain fairly complete references to the literature include: E. Farhi and L. Susskind, Physics Reports 74, 277 (1981). M.A.B. Bég and A. Sirlin, Physics Reports 88, 1 (1982).
2. See, for example, P. Fayet, Proc. of XXl Int. Conf. on High Energy Physics (Paris, 1982) p. C3-673, and references cited therein.
3. S. Dimopoulos and J. Ellis, Nucl. Phys. B 182, 505 (1981).
4. P. Fayet, ref. 2.

5. A. Salam, Proc. of XXI Int. Conf. on High Energy Physics, (Paris, 1982) p. C3-607.
6. M.A.B. Bég, H.D. Politzer and P. Ramond, Phys. Rev. Lett. 43, 1701 (1979).
   S. Dimopoulos, Nucl. Phys. B 168, 69 (1980).
7. JADE Collaboration, DESY Report No. 82-023 (1982).
   TASSO Collaboration, DESY, Report No. 82-069 (1982).
8. For example: Sau Lan Wu, Invited talk at the Eighth Irvine Conference, Irvine, California (1982).
9. Ibid., see Ref. 1.
10. A. Ali and M.A.B. Bég, Phys. Lett. B103, 376 (1981)
    J. Ellis et. al., Nucl Phys, B182, 529 (1981).
11. S. Dimopoulos and L. Susskind, Nucl. Phys. B155, 237 (1979).
    M.A.B. Bég, in: Recent Developments in High Energy Physics, Orbis Scientiae, 1980, Coral Gables, eds. B. Kursunoglu, A. Perlmutter and L.F. Scott, (Plenum, New York, 1980) p. 23.
    E. Eichten and K. Lane, Phys. Lett. B90, 125 (1980).
12. M.A.B. Bég, Rockefeller University Report No: RU82/B/35. (to be published in Phys. Lett. B).
13. M.A.B. Bég, and A. Sirlin, ref. 1.
14. M.A.B. Bég, Rockefeller University Report No: RU82/B/56.

KALUZA-KLEIN THEORIES AS A TOOL TO FIND NEW GAUGE SYMMETRIES[+]

L. Dolan

Rockefeller University

New York, N. Y. 10021

Abstract

Non-abelian Kaluza-Klein theories are studied with respect to using the invariances of multi-dimensional general relativity to investigate hidden symmetry, such as Kac-Moody Lie algebras, of the four-dimensional Yang-Mills theory. Several properties of the affine transformations on the self-dual set are identified and are used to motivate the Kaluza-Klein analysis. In this context, a system of differential equations is derived for new symmetry transformations which may be extendable to the full gauge theory.

Kaluza-Klein theories[1, 2] provide a mechanism to detect internal symmetry of a gauge theory. In particular, local gauge invariance in the four dimensional theory can be identified as a special choice of the general coordinate transformations of a multi-dimensional Einstein-Hilbert action. Recently, a new symmetry of the self-dual Yang-Mills equations has been found.[3] One may ask if this new

[+]Work supported in part by the Department of Energy under Contract Grant Number DE-AC02-83ER40033 .B000.

infinite parameter global invariance can also be derived from a symmetry of multi-dimensional gravity.[4]

In this talk, we will focus on the general coordinate invariance of general relativity. A system of differential equations is derived for a new choice of coordinate transformations, which then give infinitesimal transformations on the vector potential. These equations have complicated integrability conditions and are difficult to analyze completely. A solution can be found easily when the gauge field is a pure gauge. In general, Kaluza-Klein theories involve arbitrary gauge potentials, not just pure gauges or self-dual fields. It has been speculated that the new infinite parameter Kac-Moody-like invariance of the self-dual sector may exist in the full Yang-Mills theory. This is qualitatively suggested by the observation that 1) a similar invariance does exist in Polyakov's three dimensional loop space formulation of Yang-Mills, 2) a certain class of Kramers-Wannier self-dual systems always have an infinite set of conserved charges, and 3) the full Kac-Moody Lie algebra appears in the dual string model, an alternative description of the non-abelian gauge theory. All of these statements make no reference to the $F_{\mu\nu} = \tilde{F}_{\mu\nu}$ self-duality restriction. Thus, if the new symmetry is derived within a Kaluza-Klein framework, it could provide a way to find explicitly the transformations of the self-dual set.

The use of Kaluza-Klein was suggested by the explicit form of the self-dual transformations. They obey an equation reminiscent of a Killing equation, where the affine connection is replaced with the vector potential, i.e. the connection on a fiber bundle.

<u>Infinitesimal Transformations of $F_{\mu\nu} = \tilde{F}_{\mu\nu}$.</u>

An infinite set of symmetry transformations $\Delta^n A_\mu$ exist such that if A_μ is a solution to $F_{\mu\nu} = \frac{1}{2}\varepsilon_{\mu\nu\alpha\beta}\tilde{F}^{\alpha\beta}$, so is $A_\mu + \Delta^n A_\mu$. Here $A_\mu = A_\mu^a \frac{\sigma^a}{2i}$, σ^a are the Pauli matrices, $\varepsilon_{1234} = 1$, and we make a change of variables in Euclidean space to two complex coordinates:

$$\sqrt{2}\, y = x_4 - ix_3 \quad , \quad \sqrt{2}\, z = x_2 - ix_1 \quad ,$$

$$\sqrt{2}\,\bar{y} = x_4 + ix_3 \quad , \quad \sqrt{2}\,\bar{z} = x_2 + ix_1 \quad . \tag{1}$$

The self-dual equations become $F_{\bar{y}z} = 0$, $F_{y\bar{z}} = 0$, $F_{\bar{y}y} + F_{z\bar{z}} = 0$, where $F_{\mu\nu} = \partial_\mu A_\nu - \partial_\nu A_\mu + [A_\mu, A_\nu]$. The infinitesimal symmetries are given by

$$\begin{aligned}
&\Delta^n A_y = D_y \Omega^n \quad , && \Delta^n A_z = -D_z \Omega^n \quad , \\
&\Delta^n A_{\bar{y}} = -D_{\bar{y}} \Omega^n && \Delta^n A_{\bar{z}} = D_{\bar{z}} \Omega^n \quad ,
\end{aligned} \tag{2}$$

$$\Omega^n = -\frac{1}{2}\,(\bar{D}^{-1}\Lambda^n \bar{D} + D^{-1}\Lambda^{n\dagger} D),$$

$$\Lambda^\circ = \frac{\sigma^a}{2i}\,\rho_a \quad ,$$

$$\rho_a = \text{infinitesimal constants},$$

$$\partial_z \Lambda^{n+1} = \partial_y \Lambda^n + [J^{-1}\partial_y J, \Lambda^n] \; ,$$

$$D(x) = P\,\exp\int_{-\infty}^{x} \{d\bar{z}'\, A_{\bar{z}} + dy'\, A_y\}$$

$$\bar{D}\,(x) = [D^\dagger(x)]^{-1}, \tag{3}$$

and $D_\mu \equiv \partial_\mu + [A_\mu \; ,.$

The infinitesimal transformations of (2) satisfy the background gauge condition: $D_\mu \Delta^n A^\mu = 0$. Since $\Delta^n A_\mu$ is a symmetry of $F_{\mu\nu} = \tilde{F}_{\mu\nu}$ and $\Delta F_{\mu\nu} = D_\mu \Delta A_\nu - D_\nu \Delta A_\mu$, then $\bar{\sigma}^{\mu\nu} \Delta^n F_{\mu\nu} = 0$ and

$$\begin{aligned}
\bar{\sigma}^{\mu\nu} D_\mu \Delta^n A_\nu &= \frac{1}{2i}\,(\alpha^\mu \bar{\alpha}^\nu - \delta^{\mu\nu})\; D_\mu \Delta^n A_\nu \\
&= \frac{1}{2i}\;\alpha^\mu \bar{\alpha}^\nu\; D_\mu \Delta^n A_\nu \\
&= 0 \quad .
\end{aligned} \tag{4}$$

Here $\bar{\sigma}^{\mu\nu}$ is the antisymmetric, anti-self-dual tensor $\bar{\sigma}_{\mu\nu} = \frac{1}{2i}(\alpha^\mu \bar{\alpha}^\nu - \delta^{\mu\nu})$, where $\alpha^\mu = (-i\sigma^a, I)$, $\bar{\alpha}^\mu = (i\sigma^a, I)$ or in the complex coordinates

$$\begin{aligned}
&\alpha^4 = \bar{\alpha}^{\bar{y}} = \frac{1}{\sqrt{2}}\,(I - \sigma^3) \quad , && \alpha^z = -\bar{\alpha}^z = -\frac{1}{\sqrt{2}}\,(\sigma^1 + i\sigma^2), \\
&\alpha^4 = \bar{\alpha}^y = \frac{1}{\sqrt{2}}\,(I + \sigma^3) \quad , && \alpha^{\bar{z}} = -\bar{\alpha}^{\bar{z}} = -\frac{1}{\sqrt{2}}(\sigma^1 - i\sigma^2).
\end{aligned} \tag{5}$$

Define $f_y = D_y\Omega^n$, $f_{\bar{y}} = D_{\bar{y}}\Omega^n$, $f_z = -D_z\Omega^n$, $f_{\bar{z}} = -D_{\bar{z}}\Omega^n$. Then,

$$\alpha^{\mu}\bar{\alpha}^{\nu}(D_\mu f_\nu + D_\nu f_\mu - \frac{1}{2}g_{\mu\nu}D\cdot f) = 0. \tag{6}$$

Eq. (6) is reminiscent of the "conformal Killing equation [see Eq. (27)] where the covariant derivative with the affine connection which would appear for curved space is replaced by the gauge covariant derivative $\partial_\mu + [A_\mu\ ,$.

Non-Abelian Kaluza-Klein Theory

Kaluza-Klein theory starts from a multidimensional Einstein-Hilbert action

$$I = \int d^{4+k}z \quad \sqrt{|g(z)|} \quad R(z). \tag{7}$$

In any number of dimensions, the fundamental field is the symmetric metric tensor $g_{\mu\nu}(z)$ and $g(z) = \det g_{\mu\nu}(z)$. $R(z)$ is the scalar curvature $R(z) = g^{\mu\nu}R_{\mu\nu}$, the Ricci tensor is

$$R_{\mu\nu} = \partial_\nu\Gamma^{\rho}_{\mu\rho} - \partial_\rho\Gamma^{\rho}_{\mu\nu} + \Gamma^{\rho}_{\mu\sigma}\Gamma^{\sigma}_{\nu\rho} - \Gamma^{\rho}_{\mu\nu}\Gamma^{\sigma}_{\rho\sigma} \quad , \tag{8}$$

and the affine connection is defined to be the Christoffel symbols

$$\Gamma^{\rho}_{\mu\nu} \equiv \frac{1}{2} g^{\rho\sigma}(\partial_\mu g_{\sigma\nu} + \partial_\nu g_{\sigma\mu} - \partial_\sigma g_{\mu\nu}). \tag{9}$$

(We will be able to work in a coordinate basis.)

Eq. (7) is invariant under general coordinate transformations: for arbitrary infinitesimal transformations $f^\mu(z)$,

$$z^{\mu'} \simeq z^\mu + f^\mu(z). \tag{10}$$

Then

$$\Delta g_{\mu\nu} = f_{\mu;\nu} + f_{\nu;\mu}. \tag{11}$$

The covariant derivative is given by $f_{\mu;\nu} \equiv \partial_\nu f_\mu - \Gamma^{\alpha}_{\mu\nu}f_\alpha$. Since g is a tensor density, $g(z) \to \frac{1}{D^2}\, g(z')$ where $D = \det \dfrac{\partial z^\mu}{\partial z'^\nu} = 1-\partial\cdot f$.

Therefore $\Delta\sqrt{|g|} = \partial_\mu(f^\mu\sqrt{|g|}\ )$. The scalar R has $\Delta R = f^\mu\partial_\mu R$. Under general coordinate transformations,

$$\Delta(\sqrt{|g|}\ R) = \partial_\mu(f^\mu\sqrt{|g|}\ R). \tag{12}$$

For simplicity, we will restrict the discussion to the study of the SU(2) four-dimensional gauge theory, so that the internal dimensions K = 2 or 3, depending on whether the K-dimensional "internal" space $d^K y$ is 1) the manifold of the coset space $\frac{SU(2)}{U(1)}$ or 2) the group manifold of SU(2) itself, respectively. Each of these spaces has three Killing vectors $K_a^\alpha(y)$, a=1,2,3. (Let $z^\mu = \{x^m, y^\alpha\}$ where $\mu=1 \ldots 4+K$, $m=1,\ldots 4$, $\alpha=1, \ldots K$. The variables y^α parameterize the internal manifold which is assigned a metric $\bar{\gamma}^{\alpha\beta}(y)$.) The Killing vectors are three infinitesimal transformations on $y^\alpha \to y^\alpha + CK^\alpha{}_a(y)$ which close the SU(2) algebra. They are defined as solutions to

$$K^\alpha{}_a \partial_\alpha K^\beta{}_b - K^\alpha{}_b \partial_\alpha K^\beta{}_a = \frac{1}{C} \varepsilon_{abc} K^\beta{}_c \tag{13}$$

and

$$K_{\alpha a;\beta} + K_{\beta a;\alpha} = 0. \tag{14}$$

C is a constant which has dimensions of length. The covariant derivative of (14) is with respect to the metric $\bar{\gamma}^{\alpha\beta}(y)$. The metric $\bar{\gamma}^{\alpha\beta}$ is restricted by the requirement that solutions to (13) and (14) exist. One choice is the "natural" metric $\bar{\gamma}^{\alpha\beta} = K^\alpha{}_a K^\beta{}_a$. Eq. (14) is then automatically satisfied for any $K_{\alpha a}$. For either K, $\bar{\gamma}^\alpha{}_\beta = \delta^\alpha{}_\beta = K^\alpha{}_a K_{\beta a}$. If K=3, then $K^\alpha{}_a K_{\alpha b} = \delta_{ab}$ since $K^\alpha{}_a = \bar{\gamma}^{\alpha\beta} K_{\beta a} = K^\alpha{}_b (K^\beta{}_b K_{\beta a})$ and $K^\alpha{}_a$ is square.

A non-abelian Kaluza-Klein ansatz is

$$g_{r\nu}(z) = \begin{bmatrix} \bar{g}_{mn} - C^2 A^a_m(x) A^b_n h_{ab}(y) & CK^\alpha{}_a(y) A^a_m(x) \\ CK^\beta{}_a(y) A^a_m(x) & -\bar{\gamma}_{\alpha\beta}(y) \end{bmatrix} . \tag{15}$$

Here $h_{ab} \equiv K^\alpha{}_a(y) K_{\alpha b}(y)$ and $\bar{g}_{mn}(x)$ is the metric of the four dimensional space-time manifold. Substitution of (15) into (7) using (8) and (9) gives

$$\sqrt{|g|}R = \sqrt{|g(x)|}\sqrt{|\bar{\gamma}(y)|}\ \{\bar{R}(x) - \frac{c^2}{4} F^a_{mn}(x) F^{mnb}(x) h_{ab}(y) - \tilde{R}(y)\}. \tag{16}$$

(see Appendix A).

$\bar{R}$ is the four-dimensional curvature tensor of $\bar{g}_{mn}(x)$, $\tilde{R}$ is the K-dimensional curvature tensor of $\bar{\gamma}_{\alpha\beta}(y)$, and $F^a_{mn} = \partial_m A^a_n - \partial_n A^a_m + \varepsilon_{abc} A^b_m A^c_n$. For K=3, $h_{ab}=\delta_{ab}$ and the Einstein-Hilbert Langrangian density itself is related to the four-dimensional Yang-Mills theory:

$$\sqrt{|g|}R = \sqrt{|\gamma|}\ \{\sqrt{|\bar{g}|}\ (\bar{R} + c^2 L_{YM}) - \sqrt{|g|}\ \tilde{R}\}. \tag{17}$$

Under arbitrary coordinate transformations $z^\mu \to z^\mu + f^\mu(z)$, then $x^m \to x^m + f^m(x,y)$ and $y^\alpha \to y^\alpha + f^\alpha(x,y)$. Note that $f_\mu \equiv g_{\mu\nu} f^\nu$. Since $\Delta g_{\mu\nu} = f_{\mu:\nu} + f_{\nu:\mu}$, Eq. (15) implies

$$\Delta\bar{\gamma}_{\alpha\beta} = -(\partial_\alpha f_\beta + \partial_\beta f_\alpha - 2\tilde{\Gamma}_{\alpha\beta}{}^\gamma f_\gamma) \quad , \tag{18a}$$

$$\begin{aligned}\Delta\bar{g}_{mn} &= \partial_m \bar{f}_n + \partial_n \bar{f}_m - 2\bar{\Gamma}_{mn}{}^k \bar{f}_n \\ &\quad + CK^\alpha_a A^a_m \partial_\alpha \bar{f}_n + CK^\alpha_a A^a_n \partial_\alpha \bar{f}_m \quad ,\end{aligned} \tag{18b}$$

$$\begin{aligned}\Delta(CK_{\alpha a} A^a_m) &= -CK_{\alpha a} F_m{}^{na} \bar{f}_n + \partial_\alpha \bar{f}_m + \partial_m f_\alpha \\ &\quad -CA^a_m \{\bar{\gamma}^{\beta\gamma} f_\gamma \partial_\beta K_{\alpha a} + \partial_\alpha(\bar{\gamma}^{\beta\gamma} f_\gamma) K_{\beta a}\}.\end{aligned} \tag{18c}$$

Here $\bar{f}_m \equiv f_m + CK^\alpha_a A^a_m f_\alpha$. That is, we find that only f_α and the combination $\bar{f}_m$ appear in these expressions so that instead of f_α and f_n, we can equally well consider f_α and $\bar{f}_n$ as independent functions.

To study the symmetry transformations of Yang-Mills theory in Euclidean space, we set $\bar{g}_{mn}=\delta_{mn}$ and look for particular coordinate transformations $\bar{f}_m$ and f_α which are solutions to Eq. (18 a, b) when

$$\Delta\bar{g}_{mn} = 0$$

and

$$\Delta\bar{\gamma}_{\alpha\beta} = 0. \tag{19}$$

Since any coordinate transformation is a symmetry of $\sqrt{|g|}R$, solutions to (19) imply [from (17)]

$$\Delta(\sqrt{|g|}R) = \sqrt{|g|}c^2 \Delta L_{YM} = \partial_\mu(f^\mu \sqrt{|g|}\ R). \tag{20}$$

One solution of (18a, b) and (19) is

$$\bar{f}_m(x,y) = \bar{f}_m(x) = a_m + \omega_{mn} x^n \tag{21a}$$

and

$$f_\alpha(x,y) = CK_{\alpha a}(y)\varepsilon^a(x). \tag{21b}$$

Here a_m and $\omega_{mn} = -\omega_{nm}$ are constants, and $\varepsilon^a(x)$ are arbitrary functions of x. With this choice, (18b) becomes the Killing equation for translations and O(4) rotations:

$$\Delta\bar{g}_{mn} = \partial_m \bar{f}_n(x) + \partial_n \bar{f}_m(x) = 0. \tag{22}$$

And from (18c), the associated transformation on $A^a_m(x)$ is found:

$$\Delta(K_{\alpha a} A^a_m) = -K_{\alpha a} F_m{}^{na}\bar{f}_n \tag{23a}$$

$$-CA^a_m(K^\beta{}_b \partial_\beta K_{\alpha a} + K_{\beta a}\, \partial_\alpha K^\beta{}_b)\varepsilon^b$$

$$+K_{\alpha a}\partial_m \varepsilon^a \tag{23b}$$

or

$$\Delta A^a_m = -F_m{}^{na}\bar{f}_n + \partial_m \varepsilon^a + \varepsilon_{abc} A^b_m \varepsilon^c. \tag{24}$$

Eq. (23b) follows from (13) and (14). Eq. (24) gives the standard space-time translations and rotations and the internal local gauge symmetry of SU(2). Special conformal transformations and dilations can easily be added to (24) by noting that the Weyl transformation is also a symmetry of L_{ym} in curved space:

$$\Delta_W \bar{g}_{mn} = \Lambda(x)\bar{g}_{mn}, \qquad \Delta_W A^a_m = 0 . \tag{25}$$

Therefore, since (22) is a symmetry of $\sqrt{|g|}\, L_{ym}$, an invariance of L_{ym} in flat space is given by

$$\Delta\bar{g}_{mn} + \Delta_W \bar{g}_{mn} = \partial_m \bar{f}_n(x) + \partial_n \bar{f}_m(x) + \bar{g}_{mn}\Lambda(x) = 0 \tag{26}$$

The trace of (26) implies $\Lambda = -\frac{1}{2}\partial,\ f$, so (26) is

$$\partial_m \bar{f}_n + \partial_n \bar{f}_m - \frac{1}{2}\bar{g}_{mn}\,\partial\cdot\bar{f} = 0. \tag{27}$$

Eq. (27) is the "conformal Killing equation". It has more solutions than (22) since $\partial\cdot\bar{f}$ need not be zero. They are

$$\bar{f}_n = a_n + \omega_{nm}x^m + cx_n + c_n x^2 - 2x_n c\cdot x. \tag{28}$$

Now, (28) together with (24) gives the most general known symmetry transformations of SU(2).

Note that (21b) implies:

$$y^\alpha \to y^\alpha - CK^\alpha{}_a(y)(\varepsilon^a + \bar{f}\cdot A^a). \tag{29}$$

For consistency, i.e. when we consider these transformations, (29) means that y^α must already be a function of x^m in the K-K ansatz (15). This dependence is given by[2]

$$\frac{\partial y^\alpha}{\partial x^m} = CK^\alpha{}_a(y)(\tilde{g}^{-1}\partial_m\tilde{g})^a. \tag{30}$$

The integrability condition is satisfied since $\partial_n\partial_m y^\alpha - \partial_m\partial_n y^\alpha = CK^\alpha{}_a F^a_{mn}[\tilde{g}^{-1}\partial_m\tilde{g}] = 0$. $\tilde{g}(x)$ is a group element of SU(2): $\tilde{g}=e^{\frac{i\sigma^a}{2}\lambda^a(x)}$, $\tilde{g}^{-1}\partial_m\tilde{g}$ is an element of the algebra, infinitesimally $\tilde{g}^{-1}\partial_m\tilde{g}\sim\frac{i\sigma^a}{2}\partial_m\lambda^a$, so $(\tilde{g}^{-1}\partial_m\tilde{g})^a\sim -\partial_m\lambda^a$. From (29),

$$\partial_m y^\alpha \to \partial_m y^\alpha - CK^\alpha{}_a\partial_m(\varepsilon^a + \bar{f}\cdot A^a),$$

which is consistent with (30) for $\lambda^a = -(\varepsilon^a + \bar{f}\cdot A^a)$.

New Transformations

We now look for new solutions of (18) and (19). Instead of (21), let $\bar{f}_m(x,y)$ be a function of x^n and y^α and $f_\alpha(x,y) = 0$. With this choice, (18b) becomes

$$\begin{aligned}\Delta\bar{g}_{mn} &= \partial_m f_n(x,y) + CK^\alpha{}_a A^a_m\partial_\alpha\bar{f}_n + \partial_n\bar{f}_m(x,y) + CK^\alpha{}_a A^a_n\partial_\alpha\bar{f}_m(x,y)\\ &= 0.\end{aligned} \tag{31}$$

This is a system of first order partial differential equations. Given a new solution of (31), we can find from (18c), the associated transformation on $K_{\alpha a}A^a_m$:

$$\Delta(CK_{\alpha a}A^a_m) = -CK_{\alpha a}F_m{}^{na}\bar{f}_n + \partial_\alpha\bar{f}_n\ . \tag{32}$$

Since ΔA^a_m is a function only of x^n, we must again impose some dependence of y^α on x^m. If we use (30), then for A^a_m a pure gauge in (31), Eq. (31) implies

$$\begin{aligned}\Delta\bar{g}_{mn} &= \partial_m \bar{f}_n(x,y) + (\partial_m y^\alpha)\partial_\alpha \bar{f}_n + n \leftrightarrow m \\ &= \frac{d}{dx^m}\bar{f}_n + \frac{d}{dx^n}\bar{f}_m \\ &= 0. \end{aligned} \tag{33}$$

Therefore, for pure gauges, $\partial_\alpha \bar{f}_m$ is not restricted by (31).

In general, (31) has complicated integrability conditions, and it is difficult to analyze systematically. The open questions are: 1) whether this equation has more general solutions when $A^a_m(x)$ are restricted to be self-dual potentials, since in this case, the components of the field strengths vanish in any antiself-dual plane, 2) if such solutions exist, do they imply Kac-Moody-like transformations on $A^a_m(x)$ via (32) and 3) can the self-dual restriction be relaxed in (31) to generalize the affine algebra to $D_\mu F_{\mu\nu}=0$?

In conclusion, we remark that this entire procedure of using Kaluza-Klein to derive any symmetry of gauge theories is not systematic, since the off-diagonal elements of the metric transformation $\Delta g_{\mu\nu}$ in the multidimensional gravity theory give information about the combination $\Delta(K^\alpha{}_a(y)\, A^a_m(x))$ not just ΔA^a_m. The ideas in this talk are really presented as guide in an attempt to get some information about the difficult, unanswered question of whether a hidden symmetry is present in the four-dimensional gauge theory. It may be also that such invariance does exist but has no counterpart in the higher dimensional gravity action, being broken there explicitly by terms which vanish for the Kaluza-Klein ansatz. And another possibility is that such a symmetry is reflected, but by an internal rather than a general coordinate transformation on the metric tensor.

Appendix A

With Eq. (15), we calculate the affine connection using (8).

We have tacitly assumed we are working in a coordinate basis and with this procedure we get the standard answer. (See Eq. 16). The components are ($\alpha,\beta,\gamma=1,\ldots K$; i, j, k, m, n=1,...4).

$$\Gamma_{\alpha\beta}{}^{m} = 0$$

$$\Gamma_{\alpha\beta}{}^{\gamma} = \tilde{\Gamma}_{\alpha\beta}{}^{\gamma} = \frac{1}{2}\,\bar{\gamma}^{\gamma\rho}(\partial_\alpha\bar{\gamma}_{\rho\beta} + \partial_\beta\bar{\gamma}_{\rho\alpha} - \partial_\rho\bar{\gamma}_{\alpha\beta})$$

$$\Gamma_{\alpha j}{}^{k} = \frac{C}{2}\,F_j{}^{ka}K_{\alpha a}$$

$$\Gamma_{\alpha j}{}^{\beta} = \frac{C^2}{2}\,A^{md}K^{\beta}{}_{d}K_{\alpha a}\,F_{jm}{}^{a} - CA_j^cK^{\beta}{}_{c;\alpha}$$

$$\Gamma_{ij}{}^{\gamma} = -CK^{\gamma}{}_{a}\{A^a_{i;j} + \frac{1}{2}F_{ij}{}^{a}\} - \frac{C^3}{2}\,A^{na}K^{\gamma}{}_{a}h_{bc}\{A^b_jF^c_{in} + A^b_iF_{jn}{}^{c}\}$$

$$+ C^2A_i{}^{d}K^{\alpha}{}_{d}A_j{}^{c}K^{\gamma}{}_{c;\alpha}$$

$$\Gamma_{ij}{}^{k} = \bar{\Gamma}^k_{ij} - \frac{C^2}{2}\,h_{bc}\,\{A^b_jF_i{}^{kc} + A_i{}^{b}\,F_j{}^{kc}\}$$

where

$$K^{\beta}{}_{c;\alpha} \equiv \partial_\alpha K^{\beta}{}_{c} + \tilde{\Gamma}^{\beta}_{\alpha\gamma}K^{\gamma}{}_{a}$$

and

$$\bar{\Gamma}^k_{ij} \equiv \frac{1}{2}\,\bar{g}^{km}(\partial_i\bar{g}_{mj} + \partial_j\bar{g}_{mi} - \partial_m\bar{g}_{ij})$$

and

$$A^a_{i;j} \equiv \partial_jA^a_i - \bar{\Gamma}_{ij}{}^{k}A^a_k \quad .$$

REFERENCES

1. Th. Kaluza, Sitzungsberichte der Preuss. Akad. Wiss. (1921) 966; O. Klein, Z. Phys. 37 (1926) 895. Non-abelian Kaluza-Klein theories, which are discussed in this talk, first appeared in B. DeWitt, in Lectures at 1963 Les Houches School, Relativity, Groups, and Topology, ed. B. DeWitt (New York, Gordon and Breach, 1965) p. 725. See also Ref. 2 for a review of these.

2. A. Salam and J. Strathdee, Ann. of Phys. 141, 316 (1982).

3. L. Dolan, Phys. Lett. B113, 387 (1982).

4. L. Dolan, in preparation.

ULTRA-VIOLET FINITENESS OF THE N = 4 MODEL

Stanley Mandelstam

University of California

Berkeley, CA 94720

ABSTRACT

The N = 4 model is examined in light-cone coordinates. The advantage of such coordinates is that they can easily be applied to extended supersymmetries. In a certain form of the light-cone gauge, the perturbation expansion of the N = 4 model is free of ultraviolet divergences. As a consequence, the β-function vanishes in any order of perturbation theory in any gauge.

INTRODUCTION

The analysis of extended supersymmetric models is complicated by the difficulty of obtaining a manifestly supersymmetric formalism. In this talk I should like to show that such difficulty does not exist if one uses light-cone coordinates. The Wess-Zumino algebra for such coordinates has already been quoted in a paper by Siegel and Gates; it is simpler than the co-variant algebra.[1] Superspaces for ordinary and extended supersymmetries can easily be constructed in the light-cone frame.

Extended supersymmetric models exist where N, the number of Majorana fermions, is equal to 2 or 4. The N = 4 model, originally proposed by Gliozzi, Olive and Scherk,[2] was conjectured by Gell-Mann and Schwarz[3] to be finite in any order of perturbation theory. This conjecture has been verified up to three loops by Grisaru, Rocek and Siegel and by Tarasov.[4] Related work has been performend by Ferrara and Zumino, by Sohnius and West, and by Stelle.[5] Once one has a superspace formulation of a supersymmetric theory, it often happens that cancellations which previously appeared "miraculous" occur in a natural way. The vertex renormalization in the Wess-Zumino model is a well-known example. The wave-function renormalization, and therefore the coupling-constant renormalization, is infinite in the Wess-Zumino model. The N = 4 model is a gauge theory and, since we are using a physical gauge, we have the "naive" Ward identity which relates the vertex and wave-function renormalizations. We shall thereby be able to prove that all renormalization constants are perturbatively finite in a special form of the light-cone gauge.

The difficulty in obtaining covariant extended superspace is connected with the lack of correspondence between the number of fields and the number of particles. In any supersymmetric model, the particle multiplets will form a representation of the supersymmetry algebra. In particular, the number of bosons and fermions will be equal. When going from fields to particles this correspondence is lost, and one restores the balance by adding auxiliary fields. In the N = 4 model, there is one massless vector multiplet, six scalar multiplets and four fermion multiplets. Thus, if we count each helicity state separately, there are eight boson multiplets and eight fermion multiplets. Passing from particles to fields, we add two polarization states for the bosons and double the number of fermions. We thus have ten boson multiplets and sixteen fermion multiplets. No set of auxiliary fields has been found to restore the balance. With light-cone coordinates, the number of fields and particles is the same, and no problem arises. In fact, Green and Schwarz[6] encountered no problem in obtaining a manifestly supersymmetric formulation of the ten-dimensional string; the reason was that they were using light-cone coordinates.

WESS-ZUMINO MODEL

We shall illustrate the method by first discussing the Wess-Zumino model. In light-cone coordinates,[7] the Dirac equation for the lower two components of the four-spinor does not involve the "time" coordinate $x^+ = \frac{1}{\sqrt{2}}(x^0 + x^3)$, but only the "space" coordinates x^1, x^2, $x^- = \frac{1}{\sqrt{2}}(x^0 - x^3)$. Thus the two lower coordinates can be eliminated in favor of the two upper coordinates. The kinetic Lagrangian for the two-spinors is then quadratic. The supersymmetry transformations similarly split into two groups. They are most easily expressed in terms of the fields $A = \frac{1}{\sqrt{2}}(A_1 - iA_2)$ where A_1 and A_2 are the scalar and pseudo-scalar fields, and the helicity eigenstates $\psi = \frac{1}{\sqrt{2}}(\psi_1 + i\psi_2)$ of the spinor fields. The first group of supersymmetry transformations is:

$$\delta A = i\alpha^* \psi \quad , \tag{1a}$$

$$\delta \psi = 2p^+ i\alpha A \quad , \tag{1b}$$

where α is an infinitesimal anticommuting c-number. We use momentum space for the x^- coordinate throughout; thus

$$p^+ = i \frac{\partial}{\partial x^-}, \quad (p^+)^{-1} f(x^-) = \frac{i}{2} \int dx'^- \, \varepsilon(x'^- - x^-) \, f(x'^-) . \tag{2}$$

The second group of supersymmetry transformations connects A with the lower components of ψ, and therefore implicitly with the upper components. Since the Lorentz transformation between two different light-cone frames connect upper and lower components, we may write the transformations in the second group as a commutator between the Lorentz generators and the transformations in the first group. A Lagrangian which is invariant under both Lorentz transformations and supersymmetry transformations in the first group is automatically invariant under supersymmetry transformations in the second group. It is therefore unnecessary to consider the second group explicitly.

We may write the generator of the general supersymmetry

transformations (1) as

$$\sum_{i=1}^{2} \bar{\alpha}_i Q_i, \qquad \alpha = \alpha_1 + i\alpha_2. \tag{3}$$

The Q's then have the simple anticommutation relations

$$\{Q_i, Q_j\} = 4p^+\delta_{ij}. \tag{4}$$

To construct a superspace, we note that the first-quantized model has operators x, p together with the two supersymmetry operators Q_1 and Q_2 which connect the bosons with the fermions. After second quantization, the wave-functions become fields which are functions of coordinates corresponding to x and the Q's. One requires only one coordinate corresponding to each pair of conjugate variables, so that we have four commuting c-numbers x^μ and one anticommuting θ. Our approach thus differs from the usual approach, which would have introduced a θ and a $\bar{\theta}$, and would have used constrained fields. The fields we use are unconstrained. A treatment of light-cone superspace along more conventional lines has been given by Brink, Lindgren and Nilsson, who have recently shown that their formalism can be used, in conjunction with the modified light-cone gauge used here, to prove finiteness.[8]

Corresponding to Q_1 and Q_2, we define the two operators:

$$Q_2 : \quad D = i\{\frac{\partial}{\partial\theta} - 2p^+\theta\}, \tag{5a}$$

$$Q_1 : \quad \tilde{D} = \frac{\partial}{\partial\theta} + 2p^+\theta. \tag{5b}$$

We notice that the operators do satisfy the commutation relations (4) and, in particular, that

$$D^2 = \tilde{D}^2 = 2p^+ \tag{5c}$$

The superfield is defined as follows:

$$\phi = -i(2p^+)^{-1}\psi + \theta A. \tag{6a}$$

Note that the superfield is fermionic. We could have defined a bosonic superfield, but the above choice leads to a closer correspondence between the formulas of the present model and those of the N = 4 model. It is easily checked that the operators (5), when applied to (6), do effect the transformation (1).

We can now define the conjugate field

$$\phi^{\dagger} = -i(2p^{+})^{-1}\psi^{\dagger} + \theta A^{\dagger}. \tag{6b}$$

The operator D is real, but $\tilde{D}$ is imaginary. Hence, when applying the D's to $\phi^{\dagger}$, we must make the correspondence:

$$Q_2 : D\phi^{\dagger}, \qquad Q_1 : -\tilde{D}\phi^{\dagger}. \tag{7}$$

In writing down a supersymmetric action, two factors must be borne in mind. The first is that we cannot define covariant derivatives in our present formalism, since we have only a single θ at our disposal. The two operators D and $\tilde{D}$ anticommute with one another, but they commute with themselves, of course. The second point to bear in mind is the reversal of sign that occurs when $\tilde{D}$ acts on the conjugate field $\phi^{\dagger}$. These two possible complications cancel one another if we make the following rule: Any term in the Lagrangian can only contain factors with an even number of $\tilde{D}$'s acting on ϕ and an odd number $\phi^{\dagger}$ or vice versa. The changes induced in the Lagrangian by a supersymmetry transformation will then disappear on integration.

One might regard θ as a real variable. From the point of view of invariance under x-y rotations, it is preferable not to do so, since under such rotations, θ changes by a phase factor. According to (6), the superfields are analytic functions of θ, i.e., they are polynomials but are independent of $\bar{\theta}$. The conjugation $\phi \to \phi^{\dagger}$ is defined as Hermitian conjugation of the fields but not of θ. All of our formulas will involve only θ, $\frac{\partial}{\partial\theta}$ or $\int d\theta$. It is in this sense that our superfields only involve a single θ, whereas the conventional formalism involves two variables, θ and $\bar{\theta}$, for each complex θ.

The Langrangian is as follows:

$$L = (\partial^\mu \phi^\dagger)\, \tilde{D}(\partial_\mu \phi) - m^2 \phi^\dagger \tilde{D}\phi - \frac{\sqrt{2}i}{3}\, g\{\phi(\partial_\ell\, \phi)(2p^+\phi)) - \phi^\dagger(\partial_r \phi^\dagger)(2p^+\phi^\dagger))\}$$

$$-\sqrt{2}\, gm\{(\tilde{D}\phi)(\tilde{D}\phi)\phi^\dagger + (\tilde{D}\phi^\dagger)(\tilde{D}\phi^\dagger)\phi\} + 2g^2(\tilde{D}\phi^\dagger)(\tilde{D}\phi^\dagger)\tilde{D}\{(\tilde{D}\phi)(\tilde{D}\phi)\},$$

$$[\partial_{r,\ell} = \partial_1 \pm i\partial_2]. \qquad (8)$$

This Lagrangian does satisfy the supersymmetry rule quoted above. In light-cone coordinates, Lagrangians always contain a quartic as well as a cubic term, since there are no auxiliary fields.

N = 4 MODEL

The N = 4 model has four times as many supersymmetry operators as the Wess-Zumino model; sixteen in the covariant formulation or eight in the light-cone formalism. We may divide them into four helicity-increasing elements $Q_{a\alpha}$ and four helicity-decreasing elements $Q_{b\alpha}$; α is a new four-valued index. Associated with, but not necessarily implied by, the supersymmetry invariance is a global SU(4) invariance between the indices α.[9] In addition, the N =4 model is a gauge theory with an arbitrary gauge group; the gauge symmetry is completely separate from the supersymmetry and the global SU(4) symmetry.

We may construct the supersymmetry multiplets by successive application of the helicity decreasing supersymmetry operators to the state of highest helicity, which is equal to 1. Since the supersymmetry operators anticommute, the states must be antisymmetric in the operators. We thus have

State	Helicity	
$\vert 1\rangle$	1	One vector multiplet
$Q_{b\alpha}\vert 1\rangle$	$\frac{1}{2}$	Four Majorana spinor multiplets
$Q_{b\alpha}Q_{b\beta}\vert 1\rangle$	0	Six scalar multiplets
$Q_{b\alpha}Q_{b\beta}Q_{b\gamma}\vert 1\rangle$	$-\frac{1}{2}$	
$Q_{b1}Q_{b2}Q_{b3}Q_{b4}\vert 1\rangle$	-1	

The states with helicity $-\frac{1}{2}$ and -1 complete the vector and Majorana spinor multiplets. The N = 4 model, unlike the Wess-Zumino model, is T.C.P. self-conjugate.

It is convenient to define the combinations

$$Q_{1\alpha} = Q_{a\alpha} + Q_{b\alpha}, \tag{9a}$$

$$Q_{2\alpha} = i\{Q_{a\alpha} - Q_{b\alpha}\}, \tag{9b}$$

which satisfy the anticommutation relations

$$\{Q_{i\alpha}, Q_{j\alpha}\} = 4p^{+}\delta_{ij}. \tag{10}$$

We now represent the Q's in terms of four θ's as follows:

$$Q_{2\alpha} \; : \; D_\alpha = i\left(\frac{\partial}{\partial\theta^\alpha} - 2p^{+}\theta^\alpha\right), \tag{11a}$$

$$Q_{1\alpha} \; : \; \tilde{D}_\alpha = \frac{\partial}{\partial\theta^\alpha} + 2p^{+}\theta^\alpha. \tag{11b}$$

The D's satisfy the commutation relations

$$\{D_\alpha, D_\beta\} = \{\tilde{D}_\alpha, \tilde{D}_\beta\} = 4p^{+}\delta_{\alpha\beta}, \tag{11c}$$

$$\{D_\alpha, \tilde{D}_\beta\} = 0 \tag{11d}$$

The supersymmetry condition will then be that, for each α, any term in the Lagrangian can only contain factors with an even number of $\tilde{D}$'s acting on ϕ and an odd number on $\phi^\dagger$ or vice versa.

In writing down the superfield, we start from the highest helicity field and go downwards. Thus:

$$\phi = i(2p^{+})^{-1}V + (2p^{+})^{-1}\theta^\alpha\psi_\alpha + \frac{i}{4}\,\theta^\alpha\theta^\beta\rho^\mu_\alpha A_\mu$$
$$+ \frac{1}{3!}\,\varepsilon_{\alpha\beta\gamma\delta}\theta^\alpha\theta^\beta\theta^\gamma\psi^\delta + 2ip^{+}\theta_1\theta_2\theta_3\theta_4V^\dagger. \tag{12}$$

Each field carries a gauge index which has been suppressed. The positive- and negative-helicity vector fields and the six spinless fields have been represented by the symbols V, $V^\dagger$ and A_μ. The matrix $\rho^\mu_{\alpha\beta}$ is the appropriate Clebsch-Gordan matrix for SU(4) [or SO(6) ; an explicit form has been given by Brink, Scherk and Schwartz[9]. We specify further that the ρ's satisfy the condition

$$\rho_{\alpha\beta} = -\varepsilon^{\alpha\beta\gamma\delta}\rho^*_{\gamma\delta}. \tag{13}$$

Since the supermultiplet is self-conjugate, the fields ϕ and $\phi^\dagger$ are not independent. They are related by the equation

$$\phi^\dagger = (2p^+)^{-2}\tilde{D}\phi. \tag{14}$$

where

$$\tilde{D} = \tilde{D}_1\tilde{D}_2\tilde{D}_3\tilde{D}_4. \tag{15}$$

In proving (14) we must make the use of the condition (15).

The Lagrangian for the model is as follows:

$$L = (\partial^\mu\phi)\cdot(\partial^\mu\phi) - \frac{\sqrt{2}ig}{3}\phi.\{(\partial_\ell\phi)\times(2p^+\phi)\} + \frac{\sqrt{2}ig}{3}\phi^\dagger.\{(\partial_r\phi^\dagger) \times (2p^+\phi^\dagger)\} - \frac{g^2}{64}\sum_\alpha \{(D_\alpha\phi)\times(D_\alpha\phi)\}.(2p^+)^{-2}\{D_\alpha\phi^\dagger)\times(D_\alpha\phi^\dagger).\}. \tag{16}$$

All dot and cross products refer to the gauge degree of freedom. The field $\phi^\dagger$ is regarded as a function of ϕ according to (14).

As we have mentioned, the operator $\phi^\dagger$ in (16) is to be regarded as a function of ϕ defined by (15). On inserting (15) and (16), we find

$$L = (\partial^\mu\phi)\cdot(\partial^\mu\phi) - \frac{\sqrt{2}ig}{3}\phi\cdot\{(\partial_\ell\phi)\times(2p^+\phi)\} - \frac{\sqrt{2}ig}{3}(2p^+)^{-2}\phi\cdot\prod_\alpha(2i\partial_{x^-},_{\theta\alpha})\{[(2p^+)^{-2}\partial_r\phi]\times(2p^+)^{-1}\phi\} - \frac{g^2}{64}\sum_\alpha(2i\partial_{x^-},_{\theta\alpha})(\phi\times\phi)\cdot(2p^+)^{-2}\cdot \prod_{\beta\neq\alpha}(2i\partial_{x^-},_{\theta\beta})\{(2p^+)^{-1}\phi\times(2p^+)^{-1}\phi\}, \tag{17}$$

where

$$2i\partial_{x^-},_\theta(\phi_1,\phi_2) = (2p^+\phi_1)\frac{\partial\phi_2}{\partial\theta} - \frac{\partial\phi_1}{\partial\theta}(2p^+\phi_2). \tag{18}$$

It is understood that the four anticommuting derivatives $\partial_{x^-,\theta\alpha}$ and $\partial_{x^-,\theta\beta}$ in the third and fourth terms (17) are to be written in cyclic order. The derivatives with respect to θ in (17) thus all occur combined with the symbol $\varepsilon^{\alpha\beta\gamma\delta}$ and, since this is an SU(4) invariant combination, the Lagrangian is manifestly SU(4) invariant.

It is now a straightforward matter to obtain Feynman rules; as in the Wess-Zumino model, the fields are unconstrained.

ULTRA-VIOLET FINITENESS

To prove the vanishing of the β-function, we examine a general vertex diagram. Let us consider an external three-point vertex of the form of the second term of (17). By using the identity

$$\phi_A\cdot(\partial_\ell\phi_B \times 2p^+\phi_C) + \phi_A\cdot(\partial_\ell\phi_C \times 2p^+\phi_B)$$
$$= -(\partial_\ell\phi_A)\cdot(\phi_B \times 2p^+\phi_C)+(2p^+\phi_A)(\phi_B\times\partial_\ell\phi_C) \;, \qquad (19)$$

we note that we can choose one pair of lines meeting at the vertex, including the external line, and apply the factors ∂_ℓ and $2p^+$. We can treat the other terms of (11) similarly; each external line has at least one factor of ∂_ℓ, ∂_r, $2p^+$ or $\frac{\partial}{\partial\theta}$.

We thus find that there are more powers of p on the external lines of vertex corrections than on the external lines of the bare vertex. (A factor $\frac{\partial}{\partial\theta}$ is dimensionally equivalent to a factor $p^{\frac{1}{2}}$). It might therefore be expected that the number of powers of p on the internal lines is insufficient to give a divergence. The power counting is easily performed and confims this result: the vertex corrections are finite, provided it is legitimate to employ naive power counting, with all components of p treated equally.

In the usual light-cone gauge, such power counting is not in fact permissible.[10] The reason is that ultra-violet divergences appear from the region where p^+ and $p^+p^- - \mathbf{p}^2$ are finite, while p^- and $\mathbf{p}^2$ are large. The poles in the factors $(p^+)^{-1}$ prevent us from continuing to imaginary p^o amd thus avoiding these dangerous regions.

The condition $A^+ = 0$ does not define the light-cone gauge uniquely, since it remains true under a gauge transformation which depends on x^i and x^+ but not on x^-. The ambiguity is reflected in the iε prescription in the factors $(p^+)^{-1}$. Usually one takes a principal value prescription. If we could use the prescription $(p^+)^{-1} \rightarrow (p^+ + i\varepsilon p^-)^{-1}$, there would be no difficulty in continuing the p° integration to imaginary p°. After such continuation, it is easy to see that naive power counting is valid.

It is not difficult to show that one can define a light-cone gauge with the above iε prescription. Such a "modified light-cone gauge" is inconvenient for most purposes, since it is only invariant under Lorentz transformations which leave both p^+ and p^- unchanged. For our purposes this is the best gauge to use, since the vertex functions are finite. It is also easy to prove that all n-point functions with $n \geq 3$ are finite.

For the two-point function, the above reasoning would still allow a divergent term of the form $A\mathbf{p}^2\delta_{ij} + B\mathbf{p}_i\mathbf{p}_j$. To show that such a term does not in fact occur, we use the Ward identity in the form $\Lambda^i(p, p, 0) = -\frac{\partial}{\partial p_i}\Pi(p, p)$. This version of the Ward identity is valid only if proper Green's functions involving gluons of all four polarizations are free of singularities when any of the p^+'s becomes zero; a condition which is true in the modified light-cone gauge (though not in the usual light-cone gauge). From the differential form of the Ward identity and the finiteness of Λ, we can conclude that divergent terms proportional to $\mathbf{p}^2\delta_{ij}$ or to $\mathbf{p}_i\mathbf{p}_j$ can not occur in Π. The two-point function, and the complete model are thus finite in any order of perturbation theory.

In other gauges the wave-function renormalization will generally not be finite. The divergence is a pure gauge artifact. The β-function will always vanish, however, since its vanishing is a gauge-invariant condition.

We may finally note that our method of constructing the superspace without $\bar{\theta}$'s can also be applied to the covariant approach.

For a more detailed version of the material presented here, we refer the reader to Refs. 11 (Light-cone superspace and finiteness) and 12 (Covariant superspace).

ACKNOWLEDGMENT

Research supported by the National Science Foundation under grant number PHY-81-18547.

REFERENCES

1. W. Siegel and S.J. Gates, Nucl. Phys. B189, 295 (1981).
2. F. Gliozzi, D. Olive and J. Scherk, Nucl. Phys. B122, 253 (1977).
3. M. Gell-Mann and J. Schwarz, unpublished.
4. M. Grisaru, M. Roček, and W. Siegel, Phys. Rev. Lett. 45, 1063 (1980); A. Tarasov, (to be published).
5. S. Ferrara and B. Zumino (unpublished); M. Sohnius and P. West, Phys. Lett. 100B, 245 (1981); K. Stelle, L.P.T.E.N.S. 81/24 (1981); Proceedings of the Paris Conference on High-Energy Physics (1982). The authors of the latter two references have informed me that they have succeeded in constructing a covariant N = 2 superspace, and thereby completing an alternative proof of the finiteness of the N = 4 model.
6. M.B. Green and J.H. Schwarz, Nucl. Phys. B181, 502 (1981).
7. J. Kogut and D. Soper, Phys. Rev. D1, 2901 (1970); J.D. Bjorken, J. Kogut and D. Soper, Phys. Rev. D3, 1382 (1971).
8. L. Brink, O. Lindgren and B.E.W. Nilsson, Göteborg preprint 82/21 and preprint UTTG 1-82.
9. L. Brink, J.H. Schwarz and J. Scherk, Nucl. Phys. B121, 77 (1977).
10. J.M.Cornwall, Phys. Rev. D10, 500 (1974).
11. S. Mandelstam, Nucl. Phys. B213, 149 (1983).
12. S. Mandelstam, Phys. Lett. 121B, 30 (1983).

GRAVITATION AND ELECTROMAGNETISM COVARIANT THEORIES A LA DIRAC*

G. Papini

University of Regina

Regina, Saskatchewan S4S 0A2 Canada

Abstract

A generalization of the Weyl-Dirac theory is given in which the Dirac scalar field $\beta(x)$ is complex. The electromagnetic field finds its origin in regions of space multiconnected relative to the functions $\phi = \arg \beta$, while $|\beta|$ mediates the coupling between gravity and electromagnetism. Since the electromagnetic flux is quantized, length integrability is partly restored to the theory.

1. Introduction

Weyl's attempt to unify gravitation and electromagnetism[1] is based on a generalization of the notion of parallel transport of a vector used in general relativity. In a Riemannian space, in fact, the final direction of a vector which is displaced by parallel transport around a closed loop differs in general from the initial one. Weyl's generalization requires that in addition to the direction that the length changes also.

Thus, in a Weyl geometry, if a vector has length ℓ at a point

*Research supported by the Natural Sciences and Engineering Research Council of Canada

P, after a parallel displacement δx^{μ}, ℓ changes by

$$\delta\ell = \ell\kappa_{\mu}\delta x^{\mu} , \tag{1.1}$$

where κ_{μ} represents a vector to be interpreted as the electromagnetic potential. The quantities κ_{μ} and $g_{\mu\nu}$ specify the nature of the new space entirely. By parallel displacement around a small closed loop of area $\delta S^{\mu\nu}$, the total change in the length of the vector is

$$\delta\ell = \ell f_{\mu\nu}\delta S^{\mu\nu}, \tag{1.2}$$

with

$$f_{\mu\nu} = \kappa_{\nu,\mu} - \kappa_{\mu,\nu} . \tag{1.3}$$

Standards of length are, of course, arbitrary in this theory and can be changed locally according to

$$\ell' = \lambda(x)\ \ell , \tag{1.4}$$

where λ is an arbitrary function of the coordinates. While (1.4) affects the value of κ_{μ}, which is transformed into

$$\kappa'_{\mu} = \kappa_{\mu} + (\ell n\lambda)_{,\mu} , \tag{1.5}$$

$f_{\mu\nu}$ remains unchanged. Both gravitation and electromagnetism seem, therefore, to find adequate geometrical explanation in Weyl's theory.

Equation (1.1) does, however, raise serious questions from the physical point of view. First, standards of length can be normally established experimentally. Then atomic spectra would in general be path dependent. Although this may (in ultimate analysis), produce only very small effects, the intrinsic ambiguity of the theory determined its downfall. Prof. Dirac[2] has revived interest in Weyl's attempt by introducing two distinct metrics. If one, ds_E, is not directly measurable but physically manifest through Einstein equations, while the other, ds_A, is measured by atomic apparatus, then the objections to Weyl's theory can be removed. It is indeed sufficient to assume that Weyl's geometry applies only to ds_E, while ds_A, referred to in atomic units, remains gauge-independent. Dirac also reformulates Weyl's theory by introducing the action principle

$$I_D = \int \{-\frac{1}{4} f_{\mu\nu} f^{\mu\nu} + \beta^2 {*R} + k\beta *^{\mu}\beta_{*\mu} + c\beta^4\} \sqrt{-g} \; d^4x \tag{1.6}$$

which is linear in the scalar curvature R, as in Einstein's theory.[3] In (1.6) $\beta(\dot{x})$ is a scalar field of power -1, which plays the role of a Langrange multiplier.

When the arbitrary constants k and c are set equal to 6 and 0, respectively, the field equations become

$$\beta^2(2R^{\mu\nu} - g^{\mu\nu}R) = -E^{\mu\nu} + 4g^{\mu\nu}\beta\beta^{\rho}{}_{:\rho} - 4\beta\beta^{\mu:\nu} - 2g^{\mu\nu}\beta^{\sigma}\beta_{\sigma} + 8\beta^{\mu}\beta^{\nu} \tag{1.7}$$

and

$$f^{\mu\nu}{}_{:\nu} = 0 \;, \tag{1.8}$$

where $E^{\mu\nu}$ is Maxwell's tensor.

The gauge in which Einstein equations are derived is called the Einstein gauge and corresponds to $\beta = 1$ and $\kappa_{\mu} = 0$. It is interesting to notice that when $k \neq 6$, eq. (4) becomes

$$f^{\mu\nu}{}_{:\nu} = 2(k - 6)(\beta^2\kappa^{\mu} + \beta\beta^{\mu}) \;, \tag{1.9}$$

and that the first term on the right hand side of (1.9) gives rise to the Meissner effect in London's theory of superconductivity.[4]

2. Gauge Invariance and Non-Integrability of Length

In general relativity covariance under general coordinate transformations is necessary because regions of space exist in which Galilean reference systems are not available. This just occurs whenever a gravitational field is present. By analogy one may conclude that co-covariance is required in regions of space where there is an electromagnetic field. It does not, however, follow from Weyl's theory that a breakdown of the Einstein gauge necessarily manifests itself as an electromagnetic field. It may be only assumed that the two are related. But if indeed electromagnetism finds its origin in the failure of Riemannian geometry, one should ask what are cause and nature of this failure. Weyl's theory is mute on this subject. It just introduces an extra geometrical degree of freedom and interprets it as electromagnetism.

In looking at different aspects of electromagnetism for guidance, one notices that in London's theory of superconductivity, one has in the absence of currents,

$$\kappa_\mu = \chi_{,\mu} \quad , \tag{2.1}$$

where χ is a multivalued function for a multiconnected superconductor. For a closed path linking the multiconnected region one obtains

$$\oint \kappa_\mu dx^\mu = 2\pi n \quad , \tag{2.2}$$

where n is an integer. Thus, there may be a class of theories of the Weyl-Dirac type, patterned after London's theory, to which some measure of length integrability can be restored. In these theories the breakdown of Riemannian geometry may be topological in nature and specifically caused by a physical entity. The breakdown would manifest itself through κ_μ as implied by (2.1). In generalizing Dirac theory, we will follow this lead, already hinted at by (1.6) and (1.9).

3. Covariant Theories a la Dirac

The action integral I_D is already remarkably similar to the relativistic generalization of the Landau-Ginsburg action[5]

$$I_{LG} = \int \{ -\frac{1}{4} f_{\mu\nu} f^{\mu\nu} + \frac{1}{2} |(\partial_\mu + ieA_\mu)\beta|^2 + c_2|\beta|^2 - c_4|\beta|^4 \} d^4x. \tag{3.1}$$

By analogy, in generalizing I_D we will consider $\beta(x)$ to be a complex function of real variables. Several actions are possible, e.g.

$$I_A = \int \{ -\frac{1}{4} f_{\mu\nu} f^{\mu\nu} + |\beta|^2 {}^*R + k\beta_{*\mu}\tilde{\beta}^{*\mu} + \lambda|\beta|^4 \} \sqrt{-g}\, d^4x, \tag{3.2}$$

where $\tilde{\beta}$ is the complex conjugate of β, or

$$I_B = \int \{ -\frac{1}{4} f_{\mu\nu} f^{\mu\nu} + \mathrm{Re}[\beta^2 {}^*R + k\beta_{*\mu}\beta^{*\mu} + \lambda\beta^4] \} \sqrt{-g}\, d^4x \tag{3.3}$$

or again

$$I_C = \int \{ -\frac{1}{4} f_{\mu\nu} f^{\mu\nu} + |\beta|^2 {}^*R + k\,|\beta_{*\mu}\beta^{*\mu}| + \lambda\,|\beta|^4 \} \sqrt{-g}\, d^4x \;, \tag{3.4}$$

where k and λ are arbitrary constants. I_A, I_B, and I_C do not exhaust all possible actions and a systematic study of them has not yet been done. In this work we will focus our attention on I_C.[6,7] By defining

$$\rho = |\beta| \;, \; \phi = \mathrm{Arg}\,\beta, \tag{3.5}$$

where ρ is of power -1 and ϕ of power zero, it is easy to see that I_C is invariant under general coordinate transformations and the transformations

$$g'_{\mu\nu} = \sigma^2(x)\, g_{\mu\nu} \;, \quad \kappa'_\mu = \kappa_\mu + (\ell n\sigma)_{,\mu} \;, \tag{3.6}$$

$$\rho' = \sigma^{-1}\rho \;, \quad \phi' = \phi .$$

The variation of I_C relative to $g_{\mu\nu}$, ρ, ϕ and κ_μ gives

$$\rho^2({}^*R + 2\lambda\rho^2) + k(A^2 + B^2)^{1/2} - [\rho(\Omega\rho^{*\mu} + \Lambda\rho\phi^{,\mu})]_{*\mu} = 0 \;, \tag{3.7}$$

$$[\rho(\Omega\rho\phi^{,\mu} - \Lambda\rho^{*\mu})]_{*\mu} = 0 \;,$$

$$\frac{1}{2}\rho^2({}^*R^{\mu\nu} + {}^*R^{\nu\mu} - g^{\mu\nu\,*}R) = \frac{1}{2}(f^{\mu\nu}f^{\nu}{}_{\beta} - \frac{1}{4}g^{\mu\nu}f^{\nu\beta}f_{\alpha\beta} + \lambda g^{\mu\nu}\rho^4) +$$

$$+ \frac{k}{2}g^{\mu\nu}(A^2 + B^2)^{1/2} - \Omega A^{\mu\nu} - \Lambda B^{\mu\nu} + (\rho\rho^{*\mu})^{*\nu} + (\rho\rho^{*\nu})^{*\mu} - 2g^{\mu\nu}(\rho\rho^{*\lambda})_{*}$$

$$\frac{1}{2}f^{\mu\nu}{}_{*\nu} = (\Omega - 6)\rho\rho^{*\mu} + \Lambda\rho^2\phi^{,\mu} \;,$$

with the definitions

$$A \equiv \rho^{*\mu}\rho_{*\mu} - \rho^2\phi^{,\mu}\phi_{,\mu} \;; \quad B \equiv 2\rho\rho^{*\mu}\phi_{,\mu} \;;$$

$$\Omega \equiv (k^2 - \Lambda^2)^{1/2} = kA(A^2 + B^2)^{-1/2} \;,$$

$$A^{\mu\nu} \equiv \rho^{*\mu}\rho^{*\nu} - \rho^2\phi^{,\mu}\phi^{,\nu} \tag{3.8}$$

$$B^{\mu\nu} \equiv \rho(\rho^{*\mu}\phi^{,\nu} + \rho^{*\nu}\phi^{,\mu}).$$

A way to simplify eqs. (3.8) is suggested by the variational principle itself. If in fact we require

$$\rho_{*\mu} - \varepsilon\rho\phi_{,\mu} = 0 \;, \tag{3.9}$$

with $\varepsilon = \pm 1$, and Ω vanish, $\Lambda = k$ and eqs. (3.8) simplify considerably. With the addition to I_C of the in-invariant constraint

$$\int \rho g^{\mu\nu} \gamma_\mu (\rho_{*\nu} - \varepsilon \rho \phi_{,\nu}) \sqrt{-g}\, d^4x \quad , \tag{3.10}$$

where γ_μ is a co-vector Lagrange multiplier of power zero, one obtains the equations

$$\rho^2 ({}^*R + 2\lambda\rho^2) + 2\varepsilon\kappa\rho^2 \phi^{,\mu} \phi_{,\mu} = 0 \quad , \tag{3.11}$$

$$(\rho\rho^{*\mu})_{*\mu} = 0 \quad , \tag{3.12}$$

$$\frac{1}{2} \rho^2 ({}^*R^{\mu\nu} + {}^*R^{\nu\mu} - g^{\mu\nu\,*}R) = \frac{1}{2}(f^{\mu\beta} f^{\nu}{}_{\beta} - \frac{1}{4} g^{\mu\nu} f^{\alpha\beta} f_{\alpha\beta} + g^{\mu\nu} \lambda \rho^4)$$

$$+ \rho^2 [\varepsilon k g^{\mu\nu} \phi^{,\alpha} \phi_{,\alpha} + 2(2 - \varepsilon k) \phi^{,\mu} \phi^{,\nu} + 2\varepsilon \phi^{*(\mu\nu)} \tag{3.13}$$

$$\frac{1}{2} f^{\mu\nu}{}_{*\nu} = 4(k - 3\varepsilon) \rho^2 \phi^{,\mu} \quad , \tag{3.14}$$

in addition to (3.9). Self-consistency also requires

$$(\rho^2 \gamma^\mu)_{*\mu} = 3(\rho\rho^{*\mu})_{*\mu} = 0 \quad , \tag{3.15}$$

which has the particular solution

$$\rho^2 \gamma^\mu = \varepsilon k \rho \rho^{*\mu} \quad , \tag{3.16}$$

already used in (3.11)-(3.14). Eq. (3.9), the constraint equation, can be written explicitly in the form

$$\kappa_\mu = -(\ell n \rho)_{,\mu} + \varepsilon \phi_{,\mu} \quad . \tag{3.17}$$

It implies that in order to have $f_{\mu\nu} \neq 0$, at least on of ρ and ϕ must be multivalued. Thus, (3.17) plays the role of eq. (2.1) in London's theory. Now $\ell n \rho$ is certainly nonanalytic, but it could quite conceivably be twice continuously differentiable in most of space. In this instance we would have for any path Γ linking a multi-connected region of space

$$\oint_\Gamma \kappa_\mu dx^\mu = \varepsilon \oint_\Gamma d\phi = 2\pi n \quad , \tag{3.18}$$

where the period of the integral has been chosen to be an integer multiple of 2π, in analogy to superconductivity. It follows from (3.17) that κ_μ is nonintegrable only if space is multiconnected relative to the family of functions representing ϕ. Eq. (3.18) also indicates that length integrability is restored in part to the theory as lengths can change only by fixed amounts. Vector lengths can be compared to any point in space by displacing vectors along paths linking multiconnected regions of space. These features were anticipated in Sect. 2 and make the introduction of two metrics unnecessary.

4. Equations of Motion

The equations of motion of a particle can be obtained by adding to I_C the terms

$$I_{C_1} = m \int |\beta| \; ds \tag{4.1}$$

and

$$I_{C_2} = e \int |\beta|^{-1} \; |\beta_{*\mu} u^\mu| \; ds \quad , \tag{4.2}$$

which are generalizations of the corresponding terms of Dirac's theory. In (4.1) and (4.2), m and e are numbers which refer to mass and charge of the patticle, respecticely. I_{C_2} would in particular give rise to the Lorentz force. We also add to I_C the term[8]

$$I_{C_3} = \int \Gamma u^\mu (\rho_{*\mu} - \varepsilon\rho\phi_{,\mu}) \; ds$$

where Γ is co-scalar of power +1. It amounts to imposing length integrability along the particle world-line and to requiring that the topology be respected during the particle motion. Variation of (4.1) to (4.3) with respect to δx^μ yields the equations

$$m \frac{\Delta(\rho u_\mu)}{ds} + m\rho_{*\mu} - \Gamma_{*\mu}\left(\frac{\Delta\rho}{ds} - \varepsilon\rho \frac{d\phi}{ds}\right) + \frac{\Delta\Gamma}{ds}(\rho_{*\mu} - \varepsilon\rho\phi_{,\mu}) - \qquad (4.4)$$

$$- \frac{\varepsilon\Gamma}{\rho}\left[\rho\phi_{,\mu}\frac{\Delta\rho}{ds} - \rho_{*\mu}\rho\frac{d\phi}{ds}\right] - \rho\Gamma f_{\mu\nu}u^\nu = e\{-\rho^{-1}\frac{\Delta}{ds}\left(\Sigma\frac{\Delta\rho}{ds}\right)\rho_{*\mu}$$

$$- \frac{\Delta}{ds}\left(\rho\Sigma\frac{d\phi}{ds}\right)\phi_{,\mu} + \Sigma\frac{\Delta\rho}{ds}f_{\mu\nu}u^\nu\}$$

and

$$\rho_{*\mu} - \varepsilon\rho\phi_{,\mu} = 0 \;, \qquad (4.5)$$

with

$$\frac{\Delta\rho}{ds} \equiv \rho_{*\mu}u^\mu$$

and

$$\Sigma \equiv \left[\left(\frac{\Delta\rho}{ds}\right)^2 + \left(\rho\frac{d\phi}{ds}\right)^2\right]^{-1/2} \quad .$$

Using (4.5) in (4.4), we obtain

$$\rho_{*\mu} + \frac{\Delta(\rho u_\mu)}{ds} = \frac{1}{m}\left(\rho\Gamma + \frac{\varepsilon e}{\sqrt{2}}\right)f_{\mu\nu}u^\nu \; . \qquad (4.6)$$

The constraint alone is therefore sufficient to reproduce the entire Lorentz force. It also generates the in-invariant charge $\rho\Gamma$. While I_{C_2} is completely equivalent to I_{C_3} when the constraint is present, this is not so in the absence of I_{C_3} as the terms on the r.h.s. of (4.3) indicate. It is indeed I_{C_3} that leads to the co-covariant equation of motion given by Dirac.[2] I_{C_2} is henceforth dropped.

Eq. (4.6) can then be rewritten in the form

$$\frac{d(\rho u_\mu)}{ds} - \rho\{^{\alpha}_{\mu\nu}\}u_\alpha u^\nu + \rho_{,\mu} = \frac{1}{m}(\rho\Gamma)f_{\mu\nu}u^\nu \quad . \tag{4.7}$$

It reduces to the corresponding equation of the Einstein-Maxwell theory when $\rho = 1$. We will assume that when $g_{\mu\nu} \to \omega^2 g_{\mu\nu}$, $\Gamma \to e\omega$.

5. Conformally Flat Approximation. The Equations of Motion.

In conformally flat space $g_{\mu\nu} = \omega^2 \eta_{\mu\nu}$. On using eq. (3.12) in (3.13) and (3.14), we obtain, respectively,

$$\phi_{,\mu}{}^{\mu} + 2\phi^{,\mu}(-\kappa_\mu + \varepsilon\phi_{,\mu}) = 0 \quad , \tag{5.1}$$

$$- \rho^2[\frac{2}{\rho}\rho_{,\mu\nu} - \frac{4}{\rho^2}\rho_{,\mu}\rho_{,\nu} + \frac{1}{\rho^2}\eta_{\mu\nu}\rho^{,\lambda}\rho_{,\lambda} - \frac{2}{\rho}\eta_{\mu\nu}\rho_{,\lambda}{}^{\lambda} +$$

$$+ \frac{\lambda}{2}\eta_{\mu\nu}\rho^2] = \frac{1}{2}(f_\mu{}^\beta f_{\nu\beta} - \frac{1}{4}\eta_{\mu\nu}f_{\alpha\beta}f^{\alpha\beta}) +$$

$$+ 2(k\varepsilon - 3)\rho^2(\frac{1}{2}\eta_{\mu\nu}\phi_{,\lambda}\phi^{,\lambda} - \phi_{,\mu}\phi_{,\nu}) \quad , \tag{5.2}$$

$$f^{\mu\nu}{}_{,\nu} = 4(k - 3\varepsilon)\rho^2\phi^{,\mu} \quad . \tag{5.3}$$

Eq. (3.17) remains unchanged. By writing[7]

$$\varepsilon\phi_{,\mu} = (N/4\rho^2)(\varepsilon k - 3)^{-1}v_\mu \quad , \tag{5.4}$$

$$Nv_\mu = j_\mu \quad , \tag{5.5}$$

where N is a number density, eq. (5.2) becomes

$$- \rho^2[\frac{2}{\rho}\rho_{,\mu\nu} - \frac{4}{\rho^2}\rho_{,\mu}\rho_{,\nu} + \frac{1}{\rho^2}\eta_{\mu\nu}\rho_{,\lambda}\rho^{,\lambda} \quad \frac{2}{\rho}\eta_{\mu\nu}\rho_{,\lambda}{}^{\lambda} +$$

$$+ \frac{\lambda}{2}\eta_{\mu\nu}\rho^2] = \frac{1}{2}E_{\mu\nu} + \frac{1}{2}\frac{mc^2}{Ne^2}(\frac{1}{2}\eta_{\mu\nu}j^\lambda j_\lambda - j_\mu j_\nu) \quad . \tag{5.6}$$

The second term on the r.h.s. of (5.6) represents the energy-momentum tensor of a superconductor. Moreover, the trace of (5.6), eqs. (5.3) and (3.17) and the current conservation equation can be combined in the equations

$$(\partial_\mu + i\kappa_\mu)^2\Phi - \frac{\lambda}{12}(\varepsilon k - 3)^{-1/2}|\Phi|^2\Phi = 0 \quad , \tag{5.7}$$

$$f^{\mu\nu}{}_{,\nu} = -\frac{1}{2}i(\Phi^+\Phi_{,\mu} - \Phi\Phi^+{}_{,\mu}) + \kappa_\mu\Phi^+\Phi \quad , \tag{5.8}$$

where $\Phi\Phi^+ = |\Phi|^2$, $\Phi \equiv [16(k\varepsilon - 3)]^{1/3}\, pe^{i\chi}$, and $\chi_{,\mu} = -\kappa_\mu + \varepsilon(\varepsilon k - 3)^{1/2}\phi_{,\mu}$. Eqs. (5.6), (5.7) and (5.8) indicate that superconductivity is contained in the conformally flat approximation to the theory. We now expand ρ in the neighborhood of a constant value ρ_o

$$\rho \sim \rho_o + \tilde{\rho} + \ldots \quad ,$$

and neglect all terms quadratic in any of the variables. Eqs. (5.1), (5.3) and (3.17) become

$$\phi_{,\mu}{}^{\mu} \tag{5.9}$$

$$\tilde{\rho}_{,\mu\nu} - \frac{\lambda}{4}\eta_{\mu\nu}\rho_o{}^2\tilde{\rho} = \frac{\lambda}{12}\rho_o{}^3\eta_{\mu\nu} \tag{5.10}$$

$$\kappa^{\mu,\nu}{}_{\nu} - \lambda\rho_o{}^2\kappa^\mu = [-\varepsilon\lambda\rho_o{}^2 - 4(k - 3\varepsilon)\rho_o{}^2]\phi_,{}^{\mu} \tag{5.11}$$

$$\kappa^\mu + \frac{\tilde{\rho}_,{}^{\mu}}{\rho_o} - \varepsilon\phi_,{}^{\mu} = 0 \quad . \tag{5.12}$$

The equations of motion (4.6) in conformally flat space yield

$$\frac{du^\mu}{ds} = -(\ell n\rho)_,{}^{\mu} - (\ell n\rho)_{,\alpha}u^\alpha u^\mu + \frac{e}{m}f^{\mu\nu}u_\nu \quad , \tag{5.13}$$

which show that $\ell n\rho$ acts as a pressure. For small velocities, in the static case the force due to ρ is

$$F_j = - m(\ell n\rho)_{,j} \quad , \tag{5.14}$$

and can be calculated by solving (5.10). For $\lambda > 0$, eq. (5.10) yields[9]

$$\ell n\rho \sim \ell n \frac{2}{3} \rho_o - \alpha \frac{e^{-\sqrt{\lambda}\, \rho_o r}}{r} \quad , \tag{5.15}$$

where α is an arbitrary constant. Clearly, (5.15) gives rise to a confining force to which a particle is subject in addition to the gravitational force due to $g_{\mu\nu}$, here ignored. For $\lambda = -\eta < 0$, we obtain from (5.10)

$$\ell n\rho \sim \ell n \frac{2}{3} \rho_o + \frac{A \operatorname{Cos}(\sqrt{\eta}\, \rho_o r) + B \operatorname{Sin}(\sqrt{\eta}\, \rho_o r)}{r} \tag{5.16}$$

where A and B are arbitrary constants. The force (5.14) is now periodic in r. In both cases the range of the force is $\sim (\sqrt{\lambda}\, \rho_o)^{-1}$ and is therefore of cosmological origin.

6. Conformally Flat Approximation. Electrodynamics

Eqs. (5.11) indicate that the range of the electromagnetic interaction is also of cosmological origin. If $\lambda > 0$, the square of the photon mass is negative. One has in this case a Meissner effect. Space-time does indeed acquire the properties of a superconductor and the range of the interaction corresponds to the penetration depth. If $\lambda < 0$, the photon mass is real. Eqs. (5.11) can be solved and yield a nonvanishing κ_μ, provided (5.9) has a singular solution. It is possible to parameterize the topological singularity by writing[10]

$$\phi_{,\nu\mu} - \phi_{,\mu\nu} = G^{+}_{\mu\nu}(x) \quad , \tag{6.1}$$

where

$$G^{+}_{\mu\nu}(x) = \frac{1}{2}\,\varepsilon_{\mu\nu\lambda\rho}G^{\lambda\rho}(x) \quad , \tag{6.2}$$

and

$$G^{\lambda\rho}(x) = \int \frac{\partial[y^{\lambda},y^{\rho}]}{\partial[\tau,\sigma]}\,\delta^{4}\,(x - y(\tau,\sigma))\,d\tau\,d\sigma \tag{6.3}$$

is Dirac's world sheet function. Since we wish $\phi_{,\mu}$ to be Fourier transformable, we assume it to be single-valued

$$(\partial^{\nu}\partial^{\mu} - \partial^{\mu}\partial^{\nu})\partial^{\rho}\phi = 0 \quad .$$

This and (5.9) give

$$\phi_{,\mu\alpha}{}^{\alpha} = \partial^{\nu}G^{+}_{\nu\mu}(x) \quad , \tag{6.5}$$

which can be immediately integrated. We obtain

$$\phi_{,\mu} = \int \partial^{\nu}G^{+}_{\nu\mu}\,(x')D(x - x')d^{4}x' \quad , \tag{6.6}$$

where

$$\partial_{\alpha}\partial^{\alpha}D(x) = \delta^{4}(x).$$

Conversely, from (6.6), we reobtain (6.4), provided

$$\partial^{\mu}G_{\mu\nu}(x) = 0 \quad . \tag{6.7}$$

Eq. (5.11) now can be integrated formally. The result is

$$f_{\mu\nu} = [-\lambda\varepsilon - 4(k - 3\varepsilon)]\rho_{o}{}^{2} \int G^{+}_{\mu\nu}(x')\ \Delta\ (x - x')d^{4}x', \tag{6.8}$$

where

$$(\partial_\alpha \partial^\alpha - \lambda \rho_o^{\ 2})\Delta(x) = \delta^4(x).$$

For instance a straight, infinitely long vortex (string) can be parameterized by choosing

$$y^\mu = (\tau, o, o, \sigma) \qquad -\infty < \sigma, \ \tau < \infty \quad .$$

Then

$$G^{+}_{21}(x) = G_{03}(x) = \delta(x_1)\delta(x_2) \quad , \qquad (6.9)$$

$$\phi = \tan^{-1}\frac{x_2}{x_1} \quad , \qquad (6.10)$$

and

$$f_{12} = [-\lambda - 4(k\varepsilon - 3)]\ \rho_o^{\ 2} K_o\ (\sqrt{\lambda}\ \rho_o\ \sqrt{x_1^{\ 2} + x_1^{\ 2}}\ ) \quad , \qquad (6.11)$$

where K_o is a Bessel function of imaginary argument. A large variety of topologically singular objects can be generated in similar ways.[10] In any case, irrespective of whether $\lambda \gtrless 0$, the flux is quantized according to (3.18). One must still establish a relationship between the vector potential A_μ in electrostatic units, for instance, and κ_μ. We choose

$$A_\mu = \frac{e}{2\pi c}\,\kappa_\mu \quad , \qquad (6.12)$$

which is purely classical and yields the flux quantum e/c. Passing from the Einstein action to I_c is, on the other hand, equivalent to writing

$$\frac{Ge^2}{c^4} \propto \frac{1}{\rho^2} \quad . \qquad (6.13)$$

A similar relationship is obtained by expressing A_μ as

$$A_\mu = \frac{c}{G^{1/2}} A'_\mu \quad , \qquad (6.14)$$

where A' is dimensionless. From (3.18), (6.12), and (6.14), we obtain

$$\frac{c}{G^{1/2}} \oint A'_\mu \, dx^\mu = \frac{e}{2\pi c} \oint \kappa_\mu dx^\mu$$

The integral on the l.h.s. of (6.15) has the dimensions of a length and is quantized in units of

$$\ell_o = \frac{G^{1/2} e}{c^2} \simeq 1.4 \times 10^{-34} \text{ cm} \quad . \qquad (6.16)$$

It is interesting to notice that while the flux of κ_μ is scale independent, that of A_μ is not. Therefore, in a situation in which the flux of A_μ is constant, a scaling of lengths requires a corresponding scaling of the flux quantum $^e/c$. If, in particular, a particle were to scale upward from ℓ_o to $\sim 10^{27}$ cm, the present size of the universe, then charge would have to increase by a factor $\sim 10^{61}$.

7. Concluding Remarks

In Section 2 we expressed the hope that an answer to some of the questions raised by Weyl's theory could be found. The questions centered mainly on the lack of standards of length and of a specific mechanism for the breakdown of Riemannian geometry. It now appears possible to construct a theory in which the topology of space is altered by ϕ.[11] When topological singularities of this type are present, Riemann space behaves like Weyl's space without, however, the usual complete loss of length integrability. The electromagnetic field is then a direct manifestation of topological singularities. Without them $f_{\mu\nu}$ vanishes. In this respect the theory differs from superconductivity. Space, in fact, cannot behave as a simply-connected superconductor as it needs multiconnected regions to generate a nonintegrable κ_μ. It is also evident that the adopted form (3.17) for κ_μ, when imposed along a particle world-line, is entirely capable of reproducing the "expected" behavior of the particle and to generate the particle charge as well. Equivalent interpretations of this result are that the particle tends to retain its dimensions during the motion and that the particle world-line is a flux tube.

Interesting though incomplete insights into the workings of the theory can be gained in the weak approximation and by setting $g_{\mu\nu} = \eta_{\mu\nu}$. This has been done in Sections 5 and 6. There, ρ and κ_μ have ranges determined by the cosmological constant λ and by ρ_o, while $\tilde{\rho}$ provides Yukawa-like or cyclic corrections to the usual gravitational force due to $g_{\mu\nu}$.[12] We also have from (6.11) that a string is at the same time a vortex in equilibrium against the pressure of ρ. Lengths are quantized in the theory in units of $G^{\frac{1}{2}}e/c^2$. The question how realistic and general a theory of electromagnetism we have constructed must now be faced. We have seen in fact that the photon has a real or imaginary mass, depending on the sign of λ. If the mass is real and we take $(\sqrt{\eta}\ \rho_o)^{-1} \sim 10^{27}$ cm, the present radius of the universe, the photon mass is $\sim 10^{-65}$ g, well below the most recently determined upper limit[13] of $\sim 8 \times 10^{-49}$ g. If $\lambda > 0$, we have a Meissner effect and the range of the interaction becomes the penetration depth. Moreover, we have seen that other aspects of superconductivity are contained in weak conformally flat approximation. Superconductivity, it may be argued, is only a particular type of electromagnetism, restricted by appropriate constitutive equations.[7] If, however, the penetration length is $(\sqrt{\lambda}\ \rho_o)^{-1} \sim 10^{27}$ cm, there does not seem to be a way to distinguish between the electromagnetic effects due to a small real or imaginary photon mass at sufficient distances from the source. Possibly in the neighborhood of the source, deviations from the "expected" behavior of electromagnetic fields may be significant.

The really distinctive feature of the present version of electromagnetism is flux quantization. It is an essential ingredient of the theory and does not depend on the photon mass's being real or imaginary. It should be taken seriously,[14] at least to the extent that ours is a realistic description of electromagnetic phenomena.

APPENDIX

The Co-covariant Calculus.

Co-covariant calculus is largely characterized by the geometrical behavior of the pair $(*\Gamma^{\lambda}{}_{\mu\nu}, \kappa_\mu)$, which transforms according to the following rules:

i) under a coordinate transformation, $*\Gamma^{\lambda}{}_{\mu\nu}$ behaves as an affine connection, while κ_μ behaves as a covariant vector, and

ii) under a conformal transformation, $g_{\mu\nu} \to \lambda^2{}_{\mu\nu}$, $*\Gamma_{\mu\nu}$ is invariant, while $\kappa_\mu \to \kappa_\mu + (\ell n\lambda)_{,\mu}$. We use commas and colons to denote partial and covariant differentiation, respectively.

A co-tensor of power n and of rank (i + j) is a tensor $(\phi_A) \equiv (\phi^{\alpha_1 \ldots \alpha_i}{}_{\beta_1 \ldots \beta_j})$ which transforms, under a conformal transformation, according to $\phi_A \to \lambda^n \phi_A$. For n=0, we call ϕ_A an in-tensor. We denote the behavior of ϕ_A under a coordinate transformation $x^\mu \to x^\mu(x)$ by $\phi_A \to \sigma_A{}^B(a)\phi_B$, where $a = (a^\mu{}_\nu)$ is the Jacobian matrix of the transformation. The infinitesimal generators of the representation $(\sigma_A{}^B)$ are

$$(\sigma_A{}^B)_\mu{}^\nu \equiv \left. \frac{\partial \sigma_A{}^B(a)}{\partial a^\mu{}_\nu} \right|_{a^\alpha{}_\beta = \delta^\alpha{}_\beta} .$$

From ϕ_A we may form the co-tensor of power n given by

$$\phi_{A*\nu} \equiv \phi_{A,\nu} + (\sigma_A{}^B)_\lambda{}^\mu {*\Gamma}^\lambda{}_{\mu\nu}\phi_B - n\phi_A\kappa_\nu \quad , \tag{A.1}$$

which we call the co-covariant derivative of ϕ_A. In particular, for a co-scalar ρ of power -1, we have

$$\rho_{*\mu} = \rho_{,\mu} + \rho\kappa_\mu \tag{A.2}$$

while, for the metric tensor $g_{\mu\nu}$,

$$g_{\mu\nu*\lambda} = g_{\mu\nu,\lambda} - g_{\mu\alpha}{*\Gamma}^\alpha{}_{\nu\lambda} - g_{\nu\alpha}{*\Gamma}^\alpha{}_{\mu\lambda} - 2g_{\mu\nu}\kappa_\lambda \quad . \tag{A.3}$$

We assume $g_{\mu\nu*\lambda} = 0$, thus (A.3) can be solved to give

$$*\Gamma^\lambda{}_{\mu\nu} = \{{}^{\;\lambda}_{\mu\;\nu}\} - \delta^\lambda_\mu\kappa_\nu - \delta^\lambda_\nu\kappa_\mu + g_{\mu\nu}\kappa^\lambda \quad . \tag{A.4}$$

For successive co-covariant differentiations, we find

$$\phi_{A*\lambda\rho} - \phi_{A*\rho\lambda} = -(\sigma_A{}^B)_\mu{}^\nu {*R}^\mu{}_{\nu\lambda\rho}\phi_B + n\phi_A f_{\lambda\rho} \quad , \tag{A.5}$$

where the curvature and electromagnetic tensors are

$$*R^\mu{}_{\nu\lambda\rho} = -{*\Gamma}^\mu{}_{\nu\lambda,\rho} + {*\Gamma}^\mu{}_{\nu\rho,\lambda} + {*\Gamma}^\mu{}_{\alpha\lambda}{*\Gamma}^\alpha{}_{\nu\rho} - {*\Gamma}^\mu{}_{\alpha\rho}{*\Gamma}^\alpha{}_{\nu\lambda}$$

and

$$f_{\mu\nu} = -\kappa_{\mu,\nu} + \kappa_{\nu,\mu} \quad .$$

In particular, for $\phi_A = g_{\mu\nu}$, the left-hand side of (A.5) vanishes, implying

$$*R_{\mu\nu\lambda\rho} + {*R}_{\nu\mu\lambda\rho} + 2g_{\mu\nu}f_{\lambda\rho} = 0 \quad . \tag{A.6}$$

Furthermore, the generalized Bianchi identities of a Weyl space are

$$*R^{\lambda}{}_{[\mu\nu\rho]} = 0 \quad , \tag{A.7}$$

$$*R^{\lambda}{}_{\mu[\nu\rho*\sigma]} = 0 \tag{A.8}$$

and

$$f_{[\lambda\mu*\nu]} = f_{[\lambda\mu,\nu]} = 0 \quad . \tag{A.9}$$

By twice contracting (A.8), we obtain

$$(*R^{\mu}{}_{\nu} - \frac{1}{2}\delta^{\mu}{}_{\nu}*R)_{*\mu} = 0 \quad , \tag{A.10}$$

where the generalized Ricci tensor and the scalar curvature are

$$*R^{\mu}{}_{\nu} \equiv *R^{[\lambda\mu]}{}_{\lambda\nu}$$

and

$$*R \equiv *R^{\mu}{}_{\mu} \quad ,$$

respectively. From (A.6) and (A.7), the Ricci tensor is seen to satisfy

$$\bar{R}_{\mu\nu} - \bar{R}_{\nu\mu} = 2f_{\mu\nu} .$$

Therefore, in the limit of Riemannian geometry, $\kappa_{\mu} = 0$, we recover the identities $R_{\lambda\mu\nu\rho} = R_{[\lambda\mu]\nu\rho}$ and $R_{\mu\nu} = R_{(\nu\mu)}$, as we would expect.

Finally, we remark that the co-variant derivative of a density Ω_A of power n is given by

$$\Omega_{A*\nu} = \Omega_{A,\mu} - {}^*\Gamma^{\sigma}{}_{\sigma\lambda}\Omega_A + (\sigma_A{}^B)_\lambda{}^\mu {}^*\Gamma^\lambda{}_{\mu\nu}\Omega_B - n\Omega_A\kappa_\nu \quad .$$

In particular, the co-covariant divergence of an in-vector density is equal to its ordinary divergence.

References

1. H. Weyl, Raum. Zeit, Materie, Vierte erweiterte Auflage, Springer Verlag, Berlin 1921.
2. P.A.M. Dirac, Proc. R. Soc. London 333A, 403 (1973).
3. Notations and co-covariant calculus are summarized in the Appendix.
4. F. London, Superfluids, Vol. 1, Dover Publications, New York, 1960.
5. H.B. Nielsen and P. Olesen, Nucl. Phys. 61B, 45 (1973).
6. D. Gregorash and G. Papini, N. Cim. 63B, 487 (1981).
7. D. Gregorash and G. Papini, Phys. Letts. 82A, 67 (1981).
8. D. Gregorash and G. Papini, N. Cim 70B, 259 (1982).
9. G. Papini, Proc. Third Marcel Grossman Meeting, Shanghai, 1982 (in press).
10. M. Wadati, H. Matsumoto and H. Umezawa, Phys. Rev. 18D, 520 (1982).
11. Similar results can be obtained by introducing torsion: D. Gregorash and G. Papini, N. Cim. 55B, 37 (1980); 56B, 21 (1980); 64B, 55 (1981).
12. Somewhat similar results can be derived from the scalar-tensor theory of gravitation of Y. Fujii, Gen. Rel. Grav. 6, 29 (1975).
13. A.S. Goldhaber, M.M. Nieto, L. Davis, Sc. Am. May 1976.
14. Similar views are expressed in an entirely nongravitational context by E.J. Post, Phys. Rev. 9D, 3379 (1974).

GRAVITATIONAL WAVE EXPERIMENTS*

J. Weber

University of Maryland, College Park, MD 20742

and University of California, Irvine, CA 92717

Abstract

Gravitational wave antennas have sensitivity many orders greater than procedures for direct measurement of the geometry of spacetime.

Equations of motion are obtained as exact solutions of Einstein's equations and the coupled Maxwell Einstein equations.

An outline is given of a quantum theory of gravitational radiation antennas.

For a long period the backgrounds observed during the 1969-1974 observations were not confirmed. Recent publications have reported backgrounds with single antennas and correlations of widely separated antennas similar to the earlier observations.

Introduction

Einstein's General Relativity Theory predicts that changes in spacetime geometry may propagate as gravitational waves, as indicated in Figure 1. Departures from Euclidean geometry are indicated by triangles constructed of light rays, with sum of the angles differing from π. Thus, space is flat at aa', convex at bb' and concave at cc'.

*Supported in part by the National Science Foundation

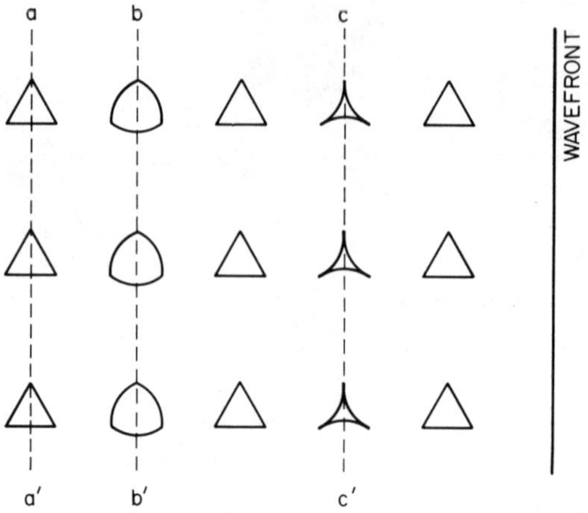

Figure 1. Propagation of Spacetime Curvature Gravitational Waves

Gravitational Wave Antennas

The spacetime curvature is measured by the Riemann tensor $R^{\mu}{}_{\alpha\beta r}$ defined in the mathematical appendix. In two dimensions there is only one component, given in terms of the angles of a light ray triangle by

$$R_{1212} = \frac{\pi - \Sigma\text{angles of a triangle}}{\text{area}} \tag{1}$$

Einstein's equations predict that gravitational waves produce alternating tidal forces, similar to the tidal forces which are well understood in Newton's law of gravitation.

Tidal forces transmitted by gravitational radiation may cause deformation of an elastic solid or displacement of the mirrors of an interferometer. For one dimensional strain (acoustic) waves, the exact equations of motion of the strain Q in an elastic solid are[2,12]

$$c^2\rho \frac{\partial Q}{\partial T^2} + F(Q,z,T,\ldots) = -c^2\rho R^z{}_{TzT}(1 + Q) \tag{2}$$

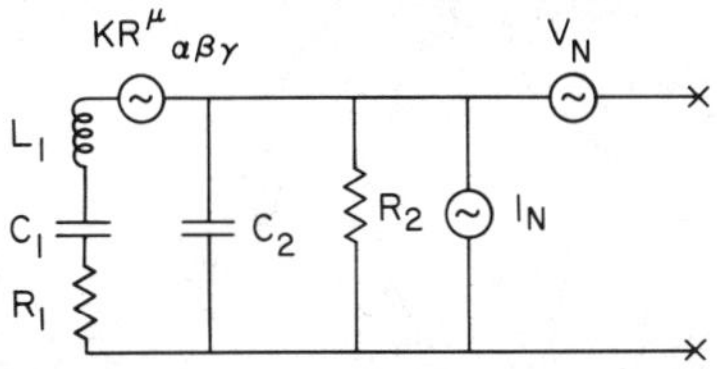

Figure 2. Solution of Maxwell Einstein Coupled Equations for Gravitational Radiation Antenna with Electromagnetic Output

In (2), T is the product of the speed of light and the time, and F is the nongravitational internal force associated with the internal damping and stress. $R^{z}{}_{TzT}$ is the component of the curvature tensor for the z direction of propagation of the strain waves,* and the "time" T.

An antenna with electromagnetic instrumentation may be described by the coupled Maxwell Einstein equations.[13] An exact solution is described by the equivalent circuit of Figure 2. This represents the input terminals of a piezoelectric crystal bonded to the elastic solid. The Riemann curvature tensor is the voltage generator driving the antenna. Resistances R_1 and R_2 may establish thermal equilibrium between the antenna normal mode and a heat bath. The noise may be computed[2] from the Einstein[3]-Nyquist[4,5,14] theory of the Brownian motion. The first successful antenna employed instrumentation of this type and is permanently displayed at the Smithsonian Institution. Figure 3 is a photograph of the antenna at the Albert Einstein exhibition of 1979.

---

*The earth may be employed as an elastic solid antenna. The noise associated[8] with quadropole modes has been measured.

Figure 3. First Gravitational Radiation Antenna at the Smithsonian Institution

It was shown first by Professor B.S. DeWitt that a single com-component of the Riemann tensor may be measured with arbitrarily great precision -- without any quantum limit.[6] Elastic solid antennas may be well isolated from terrestrial disturbances, cooled to low temperatures, and instrumented with low noise superconducting electronics.[15,16,17] The optimal length is a half acoustic wave-length at the desired resonant frequency.

Earlier, it was remarked that the tidal forces may modulate the distance between mirrors of an interferometer. This type of antenna was also considered in the earliest Maryland work.[7] There is no elastic solid acoustic wavelength limit, but there are noise problems associated with sources and mirrors.

A short pulse of gravitational radiation with wavelength λ, and curvature tensor amplitude R_0 will give a strain amplitude Q_p in an elastic solid given by

$$Q_p \sim \frac{\lambda^2 R_0}{4\pi^2} \tag{3}$$

Objects such as a pulsar would radiate continuously. A resonant antenna with quality factor Q would have strain amplitude Q_c given by

$$Q_c \sim \frac{Q\lambda^2 R_0}{4\pi^2} \tag{4}$$

The quality factor Q is given in terms of the angular frequency ω and relaxation time τ by $Q = \frac{\omega\tau}{2}$. In (3) and (4) the wavelength λ may exceed 10^7 cms. Comparison of (2) and (3) indicates that the elastic solid antenna may have observable strains for pulsed signals more than nine orders greater than the observable quantities in a direct geometrical measurement. For a quality factor of 10^9, the observable strain for continuous signals may exceed by eighteen orders the observable quantities of direct geometrical measurement.

Thorne[23] and Braginsky[24] have proposed means for reducing the noise below the levels of spontaneous emission,[11] in accordance with the DeWitt predictions. The terminals of an electromagnetically instrumented antenna provide means for coupling the late 20th century technology -- cryogenics, superconductors, computers -- to the curved spacetime of Einstein, Riemann, and Gauss.

Backgrounds

All fields of experimental physics observe backgrounds. The Maryland antennas of the period 1968-1974 observed such backgrounds.[10] Similar backgounds have been reported by the Munich Frascati group,[9] the Rome group,[19] the Rome-Maryland collaboration,[21] and the Stanford group.[20] The Rome-Maryland observations reported correlations in the output of widely separated antennas at Rome and in Maryland.

The Stanford group reported a list of events, with energy and relative frequency comparable with those observed by the Maryland antennas during the 1968-1974 period. It is, therefore, now well established that there is a background. It is not known what fraction, if any, of these background signals is due to gravitational radiation. Experiments at Maryland have shown no correlation with seismic disturbances, local cosmic ray charged particles, or with electromagnetic wave detectors at the gravitational wave antenna site. Investigation is continuing of certain low energy neutrino coherent processes.[27]

Quantum Theory of the Gravitational Radiation Antenna

Equations of motion (2) were derived from Einstein's Field Equations, following procedures pioneered by Fock[25] and Papapetrou.[26] These lead to an absorption cross section[2] of a gravitational wave antenna, σ, given approximately by

$$\sigma = \frac{15\pi G m r^2 \omega_0^2 \tau}{16c^3} \tag{5}$$

Here G is Newton's constant of gravitation, m is reduced mass,

ω_0 is the normal mode frequency, r is the separation of two point masses m giving the same quadrupole moment as the antenna, c is the speed of light, τ is the relaxation time for the antenna to reach thermal equilibrium. For kilohertz antennas under development, the cross section is smaller than $10^{-16}cm^2$. This implies that a relatively large energy flux of graviational radiation* is required even to excite a few phonons in the antenna. For this reason a semi-classical description of the ratiation is adequate. Low temperature technology provides means for observing changes in the antenna quantum state smaller than one phonon, and a quantum theory of the elastic solid antenna is required. This must describe coupling of a normal mode both to the gravitational radiation and to the many elementary particles of the antenna. The interactions of these elementary particles gives part of the noise background. We have begun with the hamiltonian

$$H = \frac{1}{2}(p^2 + \omega_0^2 q^2) + pD\sqrt{K} + mc^2 R_{0\ell 0j} r^{\ell} q^{j} \omega_0 K^{-\frac{1}{2}} + H_D + H_G \qquad (6)$$

A Riemann normal coordinate system is employed. $R_{o\ell oj}$ is the Riemann tensor of the radiation. p and q refer to the "momentum" and coordinate of the normal mode, respectively, ω_0 is again the normal mode angular frequency. m is the equivalent reduced mass and q refers to the displacement of the reduced mass. r^{ℓ} is a vector defining the equilibrium displacement of two equal bound masses having force constant K for an equivalent harmonic oscillator. H_G is the hamiltonian of the gravitational field. H_D is the hamiltonian of all elementary particles of the heat bath. D is a function of the coordinates and momenta of the heat bath. D is a function of the coordinates and momenta of the heat bath particles. The relaxation time τ is computed to be

$$\tau = 2\left\{\pi\omega_0^3 m[1-e^{-\frac{\hbar\omega_0}{kT}}]\int_0^{\infty} \rho(E_D+\hbar\omega)|\langle E_D|D|E_D+\hbar\omega\rangle|^2 \rho(E_D) f(E_D)\, dE_D\right\}^{-1} \qquad (7)$$

---

*The very small cross sections predicted by (5) imply extreme transparency[1] of most matter to gravitational radiation.

In (7) $\rho(E_D)$ is the density of states of the particles, $f(E_D)$ is the statistical weight, T is the temperature, and k is Boltzmann's constant.

For an antenna in an eigenstate, the quantum theory gives the same absorption of energy as for the antenna in thermal equilibrium This is identical with the result given by the classical theory employing Einstein's equations.

Suppose the normal mode wavefunction is Ψ, with

$$\Psi = \Sigma \; a_n(t) U_n(q) e^{-\frac{iE_n t}{\hbar}} \tag{8}$$

In (8) U_n are the harmonic oscillator energy eigenfunctions.

If the relaxation time τ is long, the antenna may be prepared in nonequilibrium states and maintained in such states for extended periods. For certain states in which the ensemble averages $\langle a_k^* a_n \rangle$ for $k \neq n$ are not zero,[22] the energy exchange is linear instead of quadratic as in the thermal equilibrium classical case. For observations over integral periods, the noise contributions may be made arbitrarily small. In agreement with DeWitt's predictions and the Thorne Braginsky results, the sensitivity of such antennas may be increased without any fundamental limit.

APPENDIX

The differential four dimensional squared interval between two events in space time is given in the special theory of relativity by

$$-ds^2 = -dT^2 + dx^2 + dy^2 + dz^2 = g_{\mu\nu} dx^\mu dx^\nu \quad . \tag{A1}$$

On the far right of (A1), the squared interval is written in a more compact form, with each repeated index a sum over three space, and one time, coordinates. The metric tensor $g_{\mu\nu}$ is written in matrix form as

$$g_{\mu\nu} = \begin{vmatrix} -1 & 0 & 0 & 0 \\ 0 & 1 & 0 & 0 \\ 0 & 0 & 1 & 0 \\ 0 & 0 & 0 & 1 \end{vmatrix}$$

In Riemannian geometry we have, again

$$-ds^2 = g_{\mu\nu}dx^{\mu}dx^{\nu} \tag{A2}$$

but now the $g_{\mu\nu}$ are functions of space and time. $g_{\mu\nu}$ is symmetric and its ten components are Einstein's gravitational potentials. Other field quantities are constructed from these potentials, by differentiation with respect to space and time coordinates, as in electromagnetic theory.

The quantities $g^{\mu\nu}$ are defined from the $g_{\mu\nu}$ by

$$g^{\mu\alpha}g_{\alpha\nu} = \delta^{\mu}{}_{\nu} \tag{A3}$$

The Christoffel symbols $\Gamma^{\mu}{}_{\alpha\beta}$ are defined by

$$\Gamma^{\mu}{}_{\alpha\beta} = \frac{1}{2} g^{\mu k} \left(\frac{\partial g_{k\alpha}}{\partial x^{\beta}} + \frac{\partial g_{k\beta}}{\partial x^{\alpha}} - \frac{\partial g_{\alpha\beta}}{\partial x^{k}} \right) \tag{A4}$$

and the Riemann Christoffel Curvature tensor is defined by

$$R^{\mu}{}_{\alpha\beta\nu} = \frac{\partial\Gamma^{\mu}{}_{\alpha\nu}}{\partial x^{\beta}} - \frac{\partial\Gamma^{\mu}{}_{\alpha\beta}}{\partial x^{\nu}} + \Gamma^{\sigma}{}_{\alpha\nu}\Gamma^{\mu}{}_{\sigma\beta} \quad \Gamma^{\sigma}{}_{\alpha\beta}\Gamma^{\mu}{}_{\sigma\nu} \tag{A5}$$

The tensor

$$R_{\mu\alpha\beta\nu} = g_{\mu k}R^{k}{}_{\alpha\beta\nu} \quad .$$

The Ricci tensor and curvature scalar are given by

$$R_{\alpha\nu} = R^{\mu}{}_{\alpha\mu\nu} \tag{A5}$$

$$R = g^{\alpha\nu}R_{\alpha\nu} \tag{A6}$$

Einstein's field equations are

$$R_{\mu\nu} - \frac{1}{2}g_{\mu\nu}R = \frac{8\pi G}{c^4} T_{\mu\nu} \tag{A7}$$

In A(7) G is Newton's constant of gravitation, c is the speed of light, $T_{\mu\nu}$ is the stress-energy of all fields in nature except the gravitational field.

References

1. J. Weber, Phys. Rev. 146, 935, 1966.
2. J. Weber, General Relativity and Gravitational Waves, Interscience-John Wiley, New York London 1961, Chapter 8; General Relativity and Gravitation, Vol. 2, A. Held (Ed.) Plenum Publishing Company 1980.
3. A. Einstein Ann der Physik 17, 549 (1905).
4. H. Nyquist Phys. Rev. 32, 110, 1928.
5. J.B. Johnson Phys. Rev. 32, 97, 1918.
6. B.S. DeWitt, In Gravitation: An Introduction to Current Research, L. Witten (Ed.) J. Wiley and Sons, New York (1962), p. 266.
7. R.L. Forward, Applied Optics 10, 2495 (1971), Reference 5.
8. J. Weber and J.V. Larson J. Geophysical Research 71, 6005, (1966) and R. Tobias thesis University of Maryland, 1978.
9. K. Billing, P. Kafka, K. Maischberger, F. Meyer, and W. Winckler Lett. Nuovo Cimento 12, 111 (1975)
10. M. Lee, D. Gretz, S. Steppel, and J. Weber, Phys. Rev. D 14, 893 (1976); J. Weber Nature 266, 5599, p. 243, March 17, 1977.
11. J. Weber Rev. Mod. Phys. 31, 3, 681, (1959).
12. J. Weber in Relativity Groups and Topology, C. DeWitt and B. DeWitt(eds.) Gordon and Breach, New York, 1964, p. 875.

13. J. Weber in Proceedings of the International School of Physics, Enrico Fermi Course LVI, B. Bertotti (ed.) Academic Press, New York, pp. 263-279.

14. R.P. Giffard Phys. Rev. D 14, 2478 (1976).

15. S.P. Boughn, W.M. Fairbank, M.S. McAshan, H.J. Paik, R.C. Taber, T.P. Bernat, D.G. Blair and W.O. Hamilton; I.A.U. Symposium #64, Boston, Massachusetts, 1974, p. 40.

16. H.J. Paik, thesis, Stanford University, HEPL 743, 1974.

17. J.P. Richard, Second Marcel Grossman Meeting Proceedings, Trieste, Italy, 1979.

18. Some of the reasons for failure of other experiments to observe the background are given in Topics in Theoretical and Experimental Gravitation Physics, V. De Sabbata and J. Weber (Editors), p. 136-137, Plenum Publishing Company, 1977.

19. Amaldi, Bonifazi, Frasca, Pallottino, Pizzella, 1980 Baltimore "Texas" Symposium on Relativistic Astrophysics.

20. Boughn, Fairbank, Giffard, Hollenhorst, Mapoles, McAshan, Michelson, Paik, Taber. Astrophysical Journal Letters 1,2, L19, October 1, 1982.

21. Ferrari, Pizzella, Lee, Weber. Physical Review D, 25, 10, 2471, May 15, 1982.

22. J. Weber, Physics Letters 81A, 9,542,23, Februrary 1981.

23. Thorne, Drever, Caves, Zimmerman, and Sandberg. Phys. Rev. Lett. 40, 667, 1978.

24. V.B. Braginsky. In Topics in Theoretical and Experimental Gravitation Physics edited by V. de Sabbata and J. Weber. Plenum Press, New York, p. 105, 1977.

25. V. Fock Revs. Modern Physics 29,3,325 (1957).

26. A. Papapetrou, Proc. Roy. Soc. A 209, 248, (1951).

27. J. Weber, Phys. Rev. D, in press.

REMARKS ON THE COSMOLOGICAL CONSTANT PROBLEM

A. Zee

University of Washington

Seattle, Washington 98195

We were asked to discuss the cosmological constant problem and will use this opportunity to present a few ideas on the subject. We should say at the outset that none of these ideas is quite satifactory and that we do not have, at the moment, any credible way of solving this nagging problem. Nevertheless, we feel that a discussion of various aspects of the problem may still be useful, and it is in this spirit of tentative gropings toward a solution that we present this talk.

Let us first explain why many physicists regard the cosmological constant problem as the biggest puzzle in contemporary physics.

Einstein's theory of gravity is, unfortunately, a two-parameter theory. Local coordinate invariance allows two terms in the gravitation action (in the second order formalism)

$$S_g = \int d^4x \sqrt{g}\ [R/G + \Lambda] \equiv S_{HE} + S_c \quad .$$

The parameters G and Λ denote Newton's constant and the cosmological constant,[1,2] respectively. (We suppress inessential factors of 4π here.) The Hilbert-Einstein action S_{HE} describes pure gravity wonderfully but the action S_c, if included with Λ having its "natural" value, spells disaster, as we will explain below.

The parameter Λ appears in Einstein's equation, giving a contribution of $\Lambda g_{\mu\nu}$ to $T_{\mu\nu}$. The cosmological constant contributes Λ to the energy density ρ and $-\Lambda$ to the pressure P. Notice that $\int d^4x\sqrt{g}$ is the volume of space-time and, thus, Λ plays a rather mysterious role. It looks sort of like a Lagrange multiplier for the volume of space-time. It is reminiscent of the role played by pressure in the dynamics of an incompressible fluid.[3]

What is the "natural" value of Λ? On dimensional grounds alone, one would expect Λ of order $1/G^2 \sim M_{p\ell}{}^4 \sim (10^{19}\ \text{GeV})^4$. Here, $M_{p\ell}$ denotes the Planck mass. This "natural" value for Λ is enormous compared to the astronomical bound on Λ. To see this, we could most easily scale up from the energy density of the Universe in photons ρ_γ. With a present cosmic background temperature $T_o \sim 3^o$ K $\sim 3\times10^{-4}$ ev, we have $\rho_\gamma = \frac{\pi^2}{15}T_o^4 \sim 10^{-50}$ GeV^4. Knowing that there are about 10^{10} photons per nucleon in the Universe, we find that the ratio of ρ_γ to the observed density in nucleons ρ_N is given by $(\rho_\gamma/\rho_N) \sim (n_\gamma/n_B)(T_o/m_N) \sim 10^{10}(3\times10^{-4}/10^9) \sim 3\times10^{-3}$. We could, therefore, state that, observationally, Λ cannot exceed ρ_γ by much more than a factor of 10^3. Thus, Λ ("observational" upper limit) $\lesssim 10^{-47}$ GeV^4. In contrast, Λ (theoretical expectation) $\sim 10^{76}$ GeV^4. Here, we have the making of the largest discrepancy between theoretical expectation and empirical observation in the history of physics:

$$\Lambda_{\text{theory}}/\Lambda_{\text{observation}} \gtrsim 10^{123}\ .$$

Of course, the value $M_{p\ell}{}^4$ for Λ is a theoretical expectation and not a prediction. Nevertheless, it is enormously puzzling that naive expectation is off by more than a hundred orders of magnitude.

Einstein, as a phenomenologist, simply set $\Lambda=0$, as one might say in modern lingo. As a classical physicist, Einstein was perfectly within his rights to do this. I do not know of any instance in his writin in which he expressed dissatisfaction with this rather arbitrary procedure. (More precisely, in classical physics, Λ has dimension mass/(length)^3 while 1/G has dimension mass/length in units in which c=1. The "natural" association $\Lambda \sim 1/\hbar G^2$ mentioned above requires the introduction of $\hbar$. However, with the advent of quantum physics, this procedure becomes highly questionable. Because of quantum fluctuations, the observed Λ is really the sum of the bare parameter one puts in the Lagrangian and a cut-off dependent correction.[4] One is entitled to set Λ to zero in the same way one sets the mass of the electron to its observed value. What bothers people is that, in physics, whenever a parameter is very small, there usually is a symmetry associated with that smallness. Thus, were the electron mass actually zero, physics would be invariant under chiral transformation on the electron field.

There is no known symmetry which would guarantee Λ to be zero. Setting Λ to zero, we do not seem to gain a new symmetry. The hope initially offered by supersymmetry and supergravity has not been realized. Indeed, it seems hard to proceed in this direction, since any symmetry one can dream up or concoct must remain unbroken to extraordinarily high orders.

The situation is exacerbated by the advent of modern particle theory, in which the concept of spontaneous symmetry-breaking plays a central role. The asymmetric vacuum and the symmetric vacuum differ, in general, in their energies. Thus, every time a symmetry breaks spontaneously, there is a heat of condensation contributing[5] to Λ. It is extremely puzzling that the parameters in the Lagrangian all have to be fine-tuned to incredible accuracy in order to insure that the net effective Λ is zero. As explained above, the units used by astronomers are such that the observation upper limit on Λ is extremely tiny when expressed in units naturally used by particle physicists. It is also important to realize that the vacuum does not break only one symmetry. We know of the breaking of chiral symmetry around 1 GeV, of electroweak symmetry around 10^2 GeV, and perhaps of grand unified symmetry around 10^{15} GeV. It is hard to imagine why the effects of this cascade of breaking should all add up just so.

The fact that gravity is not renormalizable is, in some sense, not really relevant in this context. Suppose we "start" out with the Einstein-Hilbert action S_{HE} with a bare Newton's constant G_o and suppose we cut off radiative corrections at the mass scale M_c. Then, we end up with the effective action

$$S_{eff} = \int d^4x \sqrt{g} \left(\Lambda + \frac{1}{G} R + \rho R^2 + \gamma C^2_{\mu\nu\lambda\sigma} + \ldots\right)$$

consisting of an infinite series in ascending powers of derivatives. (Here $C_{\mu\nu\lambda\sigma}$ denotes the conformal Weyl tensor.) By dimensional reasoning, we have

$$\Lambda = M_{p\ell}^{\ 4} f_o \ (M_c/M_{p\ell})$$
$$G^{-1} = G_o^{-1} + M_{p\ell}^{\ 2} f_2(M_c/M_{p\ell})$$
$$\rho = f_4(M_c/M_{p\ell})$$
$$\gamma = g_4 \ (M_c/M_{p\ell})$$

and so forth. In the effective action, the term involving 2n derivatives has a coefficient equal to $M_{p\ell}^{\ 4-2n} \ f_{2n}(M_c/ M_{p\ell})$. The "natural" expectation referred to above is that $f_{2n}(M_c/M_{p\ell})$ are all of order 1. This would lead to a phenomenologically acceptable description of classical gravity, except for the Λ term. The quantity $f_o(M_c/M_{p\ell})$ can not be of order 1; as we have seen, it has to be a number less than order 10^{-123}! Of course, mathematically, there is no reason why $f_o(M_c/M_{p\ell})$ cannot happen to be zero, perhaps for a specific value of M_c, or perhaps as $M_c/M_{p\ell} \to \infty$. However, in the presence of non-gravitational interactions and spontaneous symmetry breaking, we actually have

$$\Lambda = \Lambda_o + M_{p\ell}^{\ 4} \ f_o(M_c/M_{p\ell})$$

where we now interpret f_o to include the radiative effect of the other interactions. Setting the renormalized value of Λ equal to zero amounts to determining M_c in terms of Λ_o. The unsolved mystery is, of course, why this delicate cancellation occurs.

Since this occasion is dedicated to Dirac's eightieth birthday, we may, perhaps, draw inspiration from Dirac's well-known work on the presence of large numbers in physics. We have obtained an observational upper bound on Λ. Suppose we take this upper bound as the actual value for Λ, i.e., let $\Lambda_{obs} \sim 10^{-123}/G^2$. If so, then there is another large number in physics which Paul Dirac did not consider (perhaps rightly so, since Dirac did not include ∞ as a large number). In any case, we take

$$(G^2\Lambda_{obs})^{-1} \sim 10^{123} \quad .$$

The other large numbers in Dirac's work are

$$\alpha/(Gm_N{}^2) \sim 10^{41}$$

and $m_N\,\tau_{universe} \sim 5\times10^{41}$. Here, α denotes the fine-structure constant. The lifetime of the Universe, $\tau_{universe}$, is taken to be $\sim 10^{10}$ years. Retreating from the proposition that large numbers should not appear in physics, Dirac proposed thal all large numbers should be related and set

$$\alpha/(Gm_n{}^2) \sim m_N\tau \ .$$

We will not discuss the observational limits on Dirac's hypothesis since these have been ably reviewed[6]. It is usually assumed that α and m_N stay constant as the Universe evolves. Thus, G decreases with time as $1/\tau$.

At the risk of being accused of being numerologists, we follow Dirac and propose

$$(G^2\Lambda_{obs})^{-1} \sim (m_N\tau)^3 \quad .$$

Eliminating G, we find a formula for the cosmological constant

$$\Lambda_{obs} \sim \frac{m_N{}^3}{\alpha^2\tau} \quad .$$

Thus, Λ_{obs} decays like $1/\tau$ as the Universe ages. The ratio (Λ_{obs}/G) $\sim (m_N{}^2/\alpha)^3$ remain constaint.

Incidentally, as is well-known, there is another coincidence involving the number of nucleons in the visible Universe $N_B \sim 10^{78}$. The numerologist in us immediately recognized this to be not that far off from the square of the basic large number 10^{41}. This particular coincidence usually is stated in a different form as follows. The energy density of the Universe in nucleons ρ_N is given roughly by $N_B m_N/\tau^3$. The critical energy density ρ_c required to close the Universe is given by $1/\tau^2 \sim G\rho_c$. The ratio $\Omega \equiv \rho_N/\rho_c$ is, thus, given by

$$\Omega \sim N_B/(1/Gm_N{}^2)(m_N\tau) \quad .$$

The statement on N_B given above says that Ω is within an order of magnitude or two of being equal to 1. (By predicting Ω=1, the inflationary Universe scenario[7] may have succeeded in explaining this particular coincidence.)

We find it amusing that the number 10^{41}, its square, and its cube may all play a role in cosmology.

We hasten to add that we are well aware of the difficulties that Dirac's large number hypothesis faces. In particular, one has to replace Einstein's theory; so it is not clear what G and Λ mean. It is presumably incorrect to use simply the equation

$$\frac{\dot{R}^2}{R^2} \sim G(\rho+\Lambda)$$

with a time-varying G and Λ and, thus, it is not clear how one should discuss cosmology. Taking a cue from modern particle theory, one can readily construct theories in which fundamental parameters such as G are related to the vacuum expectation value of scalar fields and, thus, may be temperature dependent. In the context of modern field theory, the notion of varying "fundamental" parameters may be accomodated. However, in such theories, the cosmological equation is somewhat more complicated.[8] At a less fundamental level, one may also criticize the numerological aspect of Dirac's hypothesis. For instance, one might well ask why one should not use m_e in place of m_N. Nevertheless, we feel that the formula $\Lambda \sim m_N{}^3/\alpha^2\tau$ is worthy of mention.

Next, we would like to discuss an idea which Frank Wilczek and I have investigated.[9] However, as we will see, the idea is ugly in some respects and certainly sacrilegious. The basic idea is to remove $g \equiv -\det g_{\mu\nu}$ as a dynamical variable. Since Λ is the coefficient of $\sqrt{g}$ in the action, this would render Λ irrelevant to physics.

The implementation of this idea runs into a certain amount of difficulties, as we will see. We have already mentioned that it is unlikely that one can impose an additional symmetry on the Lagrangian to set Λ=0, since this symmetry must not be broken to high accuracy. Here, we consider reducing the symmetry of the theory. Suppose we reduce the local GL(4,R) symmetry of Einstein's theory to a local SL(4,R). We can say this in plain English: Instead of demanding invariance under general coordinate transformation $x^\mu \to x^{\mu'}(x)$, we reduce our demand to invariance under those coordinate transformations such that

$$\det\left(\frac{\partial x^{\mu'}}{\partial x^{\nu}}\right) = 1 \quad .$$

In that case, we are allowed to add an arbitrary function to g to go to the Langrangian. Thus, the action may be modified to read

$$S = \int d^4x \sqrt{g} \ [\frac{1}{G} (R + F(g)) + \Lambda] \quad .$$

There is no restriction on F(g). Let us choose F(g) to have a deep minimum at non-zero value of g. For instance, let $\sqrt{g}\ F(g) = \frac{\mu^2}{2}(g-1)^2$. We may naturally suppose μ to be the order of the Planck mass $M_{p\ell} \sim 1/\sqrt{G}$. If the "well" described by F(g) has very steep walls, then g is a very "stiff" variable that is largly content to take on a constant value. The dynamical variable g freezes out.

If g were a constant, then every time a symmetry is spontaneously broken by the vacuum, a new term $\int d^4x\ M^4\sqrt{g}$ would appear but would not affect physics if g is constant. Let us proceed blindly for the moment; but, as we shall see, this apparently marvelous idea will not work for an elementary reason.

Varying the action this respect to $g_{\mu\nu}$, we obtain the equation of motion

$$R_{\mu\nu} - \frac{1}{2} g_{\mu\nu}R = -GT_{\mu\nu} + f(g)\ g_{\mu\nu}$$

where we defined $f(g) \equiv [\sqrt{g}\ \ F(g)]'\ \sqrt{g}$. A possible cosmological term Λ is included in the stress-energy tensor $T_{\mu\nu}$. Taking the trace of this equation and rearranging it, we can equivalently write

$$(R_{\mu\nu} - \frac{1}{4} g_{\mu\nu}R) = -(T_{\mu\nu} - \frac{1}{4} g_{\mu\nu}T)$$

$$f(g) = T - R$$

(We suppress G from now on.)

In the limit in which $f(g) \to 0$, we recover Einstein's theory. The first of these equations states that the traceless parts of $R_{\mu\nu}$ and $T_{\mu\nu}$ are equal and opposite, while the second relates the two traces.

What is wrong with all this? Einstein gravity is a gauge theory. One is allowed to make arbitrary coordinate transformations

$x^{\mu} \to x^{\mu'}$. Four gauge conditions are necessary to fix the gauge. The addition of the term F(g) corresponds merely to the addition of a gauge-fixing term (there are still three degrees of gauge freedom). Gauge invariance is a subtle business: although the equations above look rather different from those of Einstein's theory, the physics is the same. In other words, the physics should be independent of f(g).

These general conclusions may be checked by explicit computations. Consider a simple cosmology with the flat Robertson-Walker metric

$$ds^2 = b(t)dt^2 - a(t)\ (dx^2 + dy^2 + dz^2)$$

and with $T_{\mu\nu} = \Lambda g_{\mu\nu}$. Then, the equations of motion are

$$\frac{a''}{a} - \frac{a'^2}{a^2} = \frac{a'b'}{2ab}$$

and $f(ba^3) = \Lambda - \frac{3}{4b}\frac{a'^2}{a^2}$. (Here $a' = \frac{da}{dt}$ and so forth.) We find easily the solution $a(t) \propto t^{2/3}$ and $b(t) \propto t^{-2}$. However, it is easy to see that by a coordinate transformation we can cast this solution into standard form b=1, $a \propto e^{\sqrt{(4\Lambda/3)}t}$.

In Einstein's theory, the Bianchi identity implies energy-momentum conservation $T^{\mu\nu}{}_{;\nu}=0$. Here, however, the Bianchi identity implies

$$(T^{\mu\nu} + f(g)g^{\mu\nu})_{;\nu}=0.$$

If $T^{\mu\nu}{}_{;\nu}=0$ is imposed separately (as in the simple example above with the choice $T_{\mu\nu} = \Lambda g_{\mu\nu}$), then we obtain g = constant. Conversely, if we choose a gauge in which g = constant, then $T^{\mu\nu}{}_{;\nu}=0$ follow.

It is amusing to note that, historically, Einstein actually did his first calculation[10] imposing the gauge choice g = 1. Furthermore, Einstein apparently had a great deal of difficulty mastering the fact that $g_{\mu\nu}$ is not uniquely determined by his field equations.

However, the situation is still rather strange. We thought we reduced the symmetry to SL(4,R), but then the physics turns out to be the same as that in the original GL(4,R) theory. The resolution is apparently that we have not altered the dynamical structure of the theory. With the term F(g) added the canonical momenta conjugate to $g_{\mu\nu}$ have not been changed.

The great beauty of Einstein's theory is that it fixes uniquely the interaction of matter with gravity. It almost fixes the interaction of gravity with itself (but not quite enough - precisely the cosmological constant problem). By reducing the symmetry to SL(4,R), we muck up this uniqueness completely. The invariance group local SL(4,R) is powerless in preventing one from multiplying every term in the Lagrangian by an arbitrary function of g. Thus, for example, we can write the Maxwell term as

$$\int d^4x\sqrt{g}\ F_{\mu\nu}{}^2 h_1(g)$$

or the single particle action as

$$\int ds\ h_2(g)$$

where h_1 and h_2 are arbitrary functions of g. In fact, we can also multiply the term R/G by any function of g.

If we want a truly SL(4,R) invariant theory, we have to include all these arbitrary functions of g. The physics then would be truly different from Einstein's GL(4,R) theory. We believe that, since we have an infinite number of parameters at our disposal, we should be able to fit the three classical tests. But, clearly, this program is extremely unappetizing.

A somewhat different implementation of the idea of freezing out g as a dynamical variable is to impose the condition $g = 1$. (By now, we recognize this as merely a gauge choice.)

Using Lagrange's multiplier method, we have

$$R_{\mu\nu} - \frac{1}{2} g_{\mu\nu} R = -(T_{\mu\nu} - \lambda(x) g_{\mu\nu})$$
$$\text{and} \quad g = 1 \quad .$$

(When Einstein imposed g = 1, he treated[10] just these two equations but with $\lambda(x) = 0$.) For the special case $T_{\mu\nu} = \Lambda g_{\mu\nu}$, we have $\lambda(x) = \lambda$, a constant, by virtue of Bianchi's identity. We then have the standard Einstein equation with the effective cosmological constant $(\Lambda-\lambda)$. With the metric used above, we obtain the solution

$$a = [3(\Lambda-\lambda)]^{\frac{1}{3}} t^{\frac{2}{3}} + \text{constant}$$

once again. In principle, one could choose λ to cancel Λ. As another way of saying this, let us take the trace of the equation above. This determines $\lambda(x)$. Thus, we have equivalently the equations

$$R_{\mu\nu} - \frac{1}{4} g_{\mu\nu} R = -(T_{\mu\nu} - \frac{1}{4} g_{\mu\nu} T)$$
$$g = 1$$
$$\text{and} \quad \lambda(x) = (T-R)/4$$

(These equations are to be compared with those cited a few paragraphs ago.) The cosmological constant drops out of the first equation. The first two equations determine a family of solutions. The third equation may be interpreted as determining λ given a solution or as determining the solution given . Thus, with the metric used above and with $T_{\mu\nu} = \Lambda g_{\mu\nu}$, we find that the first two equations say

$$a = K(t - t_o)^{\frac{2}{3}}$$

where K is a constant. The equation $\lambda = (T-R)/4$ determines K in in terms of λ and Λ. (At the risk of belaboring the issue, we note that we can also impose the condition $g = 1$ directly upon varying the action

$$\delta I = \int d^4x \frac{\delta L}{\delta g_{\mu\nu}} \delta g_{\mu\nu} = 0.$$

With $\delta g_{\mu\nu}$ restated to be such that $g^{\mu\nu}\delta g_{\mu\nu} = 0$, we obtain

$$\frac{\delta L}{\delta g_{\mu\nu}} = \lambda(x)\ g^{\mu\nu} \quad .$$

This, of course, amounts to a derivation of Langrange's multiplier method.

In standard applications of Lagrange's multiplier, we are used to having the multiplier λ determined. At first sight, it appears as if that is not the case here. But, in fact, in this case, there is a whole family of solutions. The choice of a specific solution determines the multiplier $\lambda(x)$. One can say arguably that, in some sense, the cosmological constant problem is solved. One simply chooses that solution for which $\lambda = \Lambda$. But, of course, there is no principle which guides us to this choice. Observational evidence is our only guide.

Incidentally, the cosmological constant has historical roots. Newton was just as perturbed as Einstein was by the fact that his equation

$$\nabla^2\phi = G\rho$$

has no (time independent) solution if space is uniformly filled by a constant ρ. Einstein[1] proposed replacing this equation by

$$(\nabla^2 + \lambda)\phi = G\rho$$

which has the solution $\phi = G\rho/\lambda$.

In connection with the cosmological constant puzzle, we must also mention the inflationary Universe scenario. In a very clever move, Guth[7] turned the cosmological constant problem around. Without inquiring as to why the cosmological constant is near zero in the present Universe, Guth simply took this fact as his observational input. After having thus "fine-tuned" the theory to give a zero cosmological constant now, one would then naturally expect the effective cosmological constant to be non-zero in the past. Thus, the Universe may have gone through an "inflationary" epoch in which the energy density of the Universe was dominated by the cosmological constant term and in which the Universe expanded exponentially. A number of long-standing puzzles were resolved in this scenario. Nevertheless, one must recognize the alternative possibility that whatever new physics, now not yet understood, responsible for setting the cosmological constant to zero now, may also have set it to zero for all time.

Does the cosmological constant problem remind us of other problems in physics? Perhaps, we can draw some inspiration.

As a first example, consider Fermi's theory of weak interaction. In its original form, the theory contains scalar, vector, tensor, axial vector and pseudoscalar interactions. On dimensional grounds alone, one would expect the coefficients of these various terms to be comparable. In this example, as we now know, the fact that some of these coefficients vanish follows from chiral symmetry. So far, no one has found the analogous symmetry in the cosmological constant problem.

As a second example, consider the fact that, in the days of the pion-nucleon theory, one could readily have written down a term $f\bar{P}e^{+}\pi^{\circ}$ in the Lagrangian which violates baryon number (Here, P, e^{+} and π° denote the proton, positron, and neutral pion fields respectively.) The "natural" theoretical expectation would have been that the dimensionless coupling f is of order 1, causing the proton to decay in $\sim 10^{-24}$ sec. The experimental lower-bound on the proton life-time is a factor 10^{61} larger, a discrepancy which almost rivals the cosmological constant discrepancy in magnitude. In present-day theory, one resolves this problem by realizing that quark fields q, rather than nucleon fields, are fundamental and so the relevant term (written schematically) $\frac{1}{M^2}\,\bar{q}\,\bar{q}\,\bar{q}\,e^+$ has operator dimension six and so has a coefficient of dimension $1/(\text{mass})^2$. The proton life-time is, thus, lengthened by a factor $(M/m_N)^4$, and the discrepancy is resolved if $M > 10^{15}$ GeV. Is it possible that, somehow, space-time has a more complicated structure than presently known, that g is not a fundamental field, and that $\sqrt{g}$ is to be replaced by a term of higher operator dimension? Again, the analogy is not perfect since one could, and did, construct a global phase symmetry forbidding baryon nonconservation. The "natural" symmetry which would forbid the cosmological constant is scale invariance. The idea of imposing scale and conformal invariances underlies the program to induce gravity. But, as is explained in our other contribution to these Proceedings, scale and conformal invariances are purely classical, and quantum fluctuations induce a cosmological constant (which, of course, may happen to be zero for reasons yet unknown).

Our third and somewhat more relevant example again refers to the Yukawa theory of strong interaction. In those days, it was mysterious why the strong interaction is parity and charge conjugation invariant and conserves strangeness. In other words, one can introduce terms such as $a\,\bar{\Lambda}\,\not{\partial}\,N$, $m_N\bar{N}(1 + b\gamma_5)N$, and so forth into the Lagrangian. (Here Λ and N denote the Λ-hyperon and nucleon fields, respectively. While some of these terms may be "rotated away" by re-defining fields, in general one gets stuck with some undesirable terms. The "natural" expectation would be that the dimensionless parameters a and b are of order unity, while, experimentally, they are known to be very small. One was at a loss to explain this. Notice that, unlike the previous example, one cannot simply impose parity, charge conjugation, and strangeness conservation on the Lagrangian and deny the existence of the problem. The point is that the weak interaction does not respect parity, charge conjugation, and strangeness conservation. The solution did not come until the formulation of quantum chromodynamics. Gauge invariance allows the offending terms to be simply transformed away. With the right theory, a nagging problem easily disappeared. We rather like the notion that this example may be telling us something. Perhaps Einstein's theory, beautiful though it is, is to the true theory of gravity what Yukawa's theory is to quantum chromodynamics. Einstein's theory has been strikingly successful as a phenomenological theory of gravitational phenomena, but attempts at interpreting it as a fundamental theory run into severe and basic difficulties. Besides the cosmological constant puzzle, we mention nonrenormalizability and ghost (See our other contribution to these Proceedings for further discussion. As explained there, Einstein's theory is also in some sense ill-matched to four-dimensional space-time.). We would not be surprised if the cosmological constant problem simply disappears in the fundamental theory of gravity.

Our fourth, and perhaps also relevant, example concerns the so-called strong CP problem in quantum chromodynamics. Because of topological effects in Yang-Mills theory, a term $\Theta F_{\mu\nu}\tilde{F}^{\mu\nu}$ may be included in the Lagrangian. Again, the "natural" theoretical expectation is that the dimensionless parameter Θ is of order 1, while, experimentally, Θ must be no larger than $\sim 10^{-10}$. The most attractive resolution of this discrepancy, namely that proposed by Peccei and Quinn, involves arranging the couplings of the Higgs fields $\phi(x)$ of the theory in such a way that Θ may be absorbed into the phases of $\phi(x)$ by performing a chiral transformation which removes the unwanted $\Theta F_{\mu\nu}\tilde{F}^{\mu\nu}$ term. Schematically, under the chiral transformation in question, a typical Higgs field $\phi(x) \to \phi(x)e^{i\Theta} = |\phi(x)|\, e^{i(\alpha(x)+\Theta)}$. Here, $\alpha(x)$ denotes the phase of $\phi(x)$. For the cosmological constant problem one might want to try scale transformation on the fields $g_{\mu\nu}(x) \to g_{\mu\nu}(\lambda x)$, $A_\mu(x) \to \lambda A_\mu(\lambda x)$, and so forth. Then, $\int d^4x \sqrt{g(x)}\, \Lambda \to \int d^4x \sqrt{g(x)}\, (\Lambda/\lambda^4)$. Of course, in the limit $\lambda \to \infty$, not only the cosmological constant but all non-dimension-four terms, such as the Einstein-Hilbert term in the Lagrangian, are scaled away. This is, of course, just the induced gravity program. In the strong CP problem, the crucial feature is that chiral transformation on fermion fields affects the $\Theta F_{\mu\nu}\tilde{F}_{\mu\nu}$ term via the axial anomaly. But there is also an anomaly associated with scale transformation. So, perhaps, this approach could work. However, $\lambda = \infty$ is a rather extreme scale transformation.

The Peccei-Quinn symmetry turns the numerical parameter Θ effectively into a dynamical field. More accurately, Θ is additively absorbed into the phase field $\alpha(x)$ degree-of-freedom contained in the Higgs field. In effect, Θ becomes the axion field, which can be made "invisible" if desired.

Perhaps, we can also somehow promote Λ into a dynamical field, $\Lambda(x)$ say. The idea is for $\phi(x)$ to find its own "natural" value, hopefully, zero. But the $\Lambda(x)$ field has to be very "clever" so that, at each stage of spontaneous symmetry breaking when a vacuum energy V_o appears, $\Lambda(x)$ "moves" to take on the value $-V_o$. So $\Lambda(x)$ must somehow know about V_o and be "coupled" to the symmetry-breaking mechanism. The reader will recognize that this picture bears a certain resemblance to our discussion of the Lagrange multiplier approach (with some suitable change of notation: $\lambda(x) \to \Lambda(x)$ and $\Lambda \to V_o$).

This discussion is tantamount to introducing a scalar field which we denote conventionally by $\phi(x)$ and which must be made to move[11] to counter any cosmological constant Λ. Since the scalar field must know about Λ, ϕ has to be coupled to the curvature, via terms such as $\phi^2 R$, $\phi^2 R^2$, $\phi^2 C^2_{\mu\nu\lambda\sigma}$, and so forth, in the Lagrangian. One then studies the resulting equations of motion, such as those written down in, for instance, ref. 8.

In another extremely interesting approach, one regards the cosmological constant as the property of the particular ground state (i.e., vacuum) of the Universe at a particular instant. For example, in induced gravity, Λ is given simply by $\langle 0|T|0\rangle$ and, thus, depends on the "vacuum state" $|0\rangle$ (T denotes the trace of the stress energy tensor as before). However, the cosmological evolution of $|0\rangle$ depends on Λ. Schematically, we may write $\frac{\partial}{\partial t}|0\rangle = F[\Lambda]$. Thus, $\frac{\partial\Lambda}{\partial t}$ is a functional of Λ. It is, thus, conceivable that, with cosmic evolution, the cosmological "constant" might relax to zero in the present Universe.[12] In other words, particle creation due to the exceptional expansion caused by Λ may act back on Λ in such a way so as to drive Λ towards zero. However, it is not clear how to carry out the back-reaction calculation correctly.

A first step in this direction has been taken recently by Parker[13] who hypothesized, arbitrarily, that the value of Λ is such that as to minimize particle production in the expanding Universe. This minimization condition amounts to imposing, in addition to Einstein's equation (with the cosmological constant), the equation

$$R = 0 \quad .$$

Tracing Einstein's equation, one finds

$$R = T + \Lambda \quad .$$

When the Universe was radiation-dominated, $T = 0$ and the condition $R = 0$ implied $\Lambda = 0$.

We will not comment on how convincing this arguemnt is. The condition $R = 0$ can be implemented by a Lagrange multiplier. Call the multiplier $\phi(x)$. Once could introduce in the Lagrangian a term such as $\phi(x)\ R(x)$ or $\phi(x)\ R^2(x)$, not allowing $\phi(x)$ to appear in any other terms. From the field theory point of view, the resulting Lagrangian looks awfully strange and is certainly not tenable as a quantum field theory. One may want to introduce other terms involving ϕ, such as a kinetic energy term $(\partial_\mu\phi)^2$. However, the resulting equation is then not $R = 0$ but some equation determining ϕ. In any case, we see that this discussion is not unrelated to the discussion in one of the preceding paragraphs.

Finally, there is also the possibility[14] that the field theory formulation of the "vacuum" is incorrect or inconvenient. The cosmological constant Λ represents the stress-energy in the vacuum. In modern particle theory, the vacuum is a roiling sea of fluctuations, topological excitations, and who-knows-what-else. Perhaps, to paraphrase Stein, a vacuum is a vacuum is a vacuum. In the standard formulation of quantum physics, in order to calculate an energy change, one must calculate the energy shift of the state in question <u>and</u> of the vacuum (i.e., ground state) and then take the difference. Feynman[14] suggested that there ought to be a formulation of field theory in which one can calculate ΔE of a state directly without referring to the ground state. It is conceivable that in this formulation, if it in fact exists, the cosmological constant may disappear of its own accord.

In conclusion, we have considered the cosmological constant problem from several points of view. We believe that some of the examples we cited, from which we hope to draw some inspiration, may turn out to be relevant. Personally, we like the idea that Einstein's theory is somehow an effective theory, as we discussed in connection with our second and third examples. In particular, think of the non-linear chiral Lagrangian of low-energy pion-nucleon physics. The theory is invariant under a non-linear chiral transformation on the pion and nucleon fields. The theory has an elegant algebraic structure; it is, however, non-renormalizable but manages to describe low energy phenomenology quite well. The meson fields are sometimes described by a unitary matrix field; $U(x)$; U and U^{-1} appear in the Lagrangian reminiscent of $g_{\mu\nu}$ and $g^{\mu\nu}$. The theory suffers from the fact that we do not understand why various terms, which will lead to assorted unobserved phenomena such as proton decay, strangeness violation and parity violation in hadron physics, are not present in the theory. We see that the situation in hadron physics was remarkably similar to the situation with Einstein gravity. We now understand the chiral theory to be an effective low-energy representation of the underlying quantum chromodynamics. The vexing problem of why terms not seen experimentally are not forbidden theoretically disappeared automatically. Hopefully, the same development may occur for gravity. Just as quarks underlie meson fields, could there by a fermionic underpinning for the metric?

Physics has invariably progressed when there is a paradoxical confrontation between a well-trusted theory and irrefutable experimental evidence. (Indeed, the lack of significant paradoxes in non-gravitational particle physics may well impede further progress.) One need only think of the ultraviolet catastrophe as an example. On a lesser scale, and more recently, the absence of strangeness changing neutral current led to the invention of charm and the discrepancy between the theoretical and experimental values of the neutral pion life-time suggested the notion of color. In contemporary physics, the biggest outstanding paradox is, without question, the cosmological constant paradox, which has been bothering physicists for over half a century. The solution to this problem will undoubtedly shed much light on the true nature of gravity.

ACKNOWLEDGEMENT

It is a pleasure to thank D. Boulware, L. Brown, R. Feynman, and F. Wilczek for clarifying discussions on various aspects of this paper.

REFERENCES

1. A. Einstein, "Cosmological Considerations on the General Theory of Relativity", Sitzungs. d. Preuss. Akad. d. Wissen. 1917.
2. A. Einstein, "Do Gravitational Fields Play an essential part in the Structure of the Elementary Particles of Matter?" Sitzungs d. Preuss. Akad. d. Wissen. 1917.
3. For example, Mechanics of Deformable Bodies, by A. Sommerfeld (English edition, Academic Press, 1950) p. 89. We thank F. Wilczek for showing us this reference.
4. Ya. B. Zel'dovich, JETP Lett. 6, 316 (1967).

5. A. Linde, JETP Lett. 19, 183 (1974); J. Dreitlein, ibid. 33, 1243 (1974); M. Veltman, Phys. Rev. Lett. 34, 777 (1975); S. Bludman and M. Ruderman, ibid. 38, 255 (1977).
6. For example, F. Dyson, in Aspects of Quantum Theory (Proceedings of a Conference to celebrate Dirac's 70th Birthday) ed. by A. Salam and E. Wigner.
7. A.H. Guth, Phys. Rev. D23, 347 (1981) and recent reviews.
8. A. Zee, Phys. Rev. Lett. 74, 417 (1979); 44, 703 (1980).
9. F. Wilczek and A. Zee, unpublished.
10. A. Einstein, Ann. der Physik 49 (1916). Note specifically the discussion around eqs. (9, 44, 47 and 47a). In particular, $R_{\mu\nu}$ simplifies tremendously.
11. F. Wilczek, private communication.
12. S. Adler, Rev. Mod. Phys. 54, 729 (1982); N.P. Myhrvold, Princeton preprint (1983).
13. L. Parker, Winsconsin-Milwaukee preprint (1983).
14. R.P. Feynman, private communication.

A NEW FORMULATION OF N=8 SUPERGRAVITY AND ITS EXTENSION TO TYPE II SUPERSTRINGS*

John H. Schwarz
California Institute of Technology
Pasadena, CA 91125

Abstract

A new extended supergravity theory in ten dimensions is formulated in terms of an unconstrained scalar light-cone superfield. Since this theory describes the massless sector of type II superstring theory, the formalism is then extended to fields that are functionals of strings superspace coordinates. In this way a field theory of superstrings is obtained. The theory is expected to be nonpolynomial, but so far it has only been constructed to order κ.

1. Introduction

Supersymmetrical string theories (or "superstring" theories) have been proposed for use as a functional theory of elementary particles unifying the description of gravitation with the other interactions.[1] This program has advanced rapidly in the last few years with the development by Michael Green and myself of a formalism in which the spacetime supersymmetry of the theories is revealed.[2] The most recent developments, described here in abridged form, include the construction of a new extended supergravity theory

*Work supported in part by the US Department of Energy under contract No. DE-ACOS-81-Er40050.

in ten dimensions that corresponds to the massless sector of type II superstring theory[3] and a field-theoretic description of the superstrings themselves based on a superfield that is a functional of light-cone superspace coordinates.[4,5]

Conventional quantum field theories based on a finite number of point-particle fields (including supergravity) appear incapable of providing a perturbatively finite or renormalizable theory of gravitation. The outlook for superstring theories is more sanguine, even though no proofs have been given yet beyond one loop. Superstring theories contain two fundamental parameters, a coupling constant κ and a Regge-slope parameter α' that is inversely proportional to the string tension. (The combination $\kappa(\alpha)^{-2}$ is dimensionless.) In addition, there is a length scale R characterizing the size of six compact dimensions, required for realism since the theories are altogether ten-dimensional. It is hoped that R will be related to α' and κ when one succeeds in finding classical solutions that describe the compactification. This step should also be responsible for super-symmetry breaking and the introduction of Yang-Mills gauge symmetries. It will become feasible to explore nontrivial compactifications when the full nonlinear field theory is explicitly constructed, but so far only the order κ interactions have been described in the appropriate field-theory language.

Both type I and type II superstring theories are free from ghosts and tachyons and require ten-dimensional spacetime for their consistency as quantum theories. Type I theories have one ten-dimensional supersymmetry and describe unoriented open and closed strings. They allow orthogonal or symplectic Yang-Mills gauge groups to be introduced and reduce to N=4 Yang-Mills field theory in a suitable α', $R \to 0$ limit. It has been shown at one loop that the only divergences occurring in the perturbation expansion of type I theories can be removed by renormalization of α'. Theory II has two ten-dimensional supersymmetries and describes oriented closed strings only. It is a unique theory, with no freedom to choose a gauge group, and reduces to N=8 supergravity in a suitable α', $R \to 0$ limit. The evidence from one-loop and topological considerations suggests that theory II is finite to all orders in perturbation theory, even though the limiting N=8 supergravity theory probably has ultraviolet divergences starting at three loops.

The purpose of this talk is to sketch the field-theoretic description of type II superstring theory. In older string theories one started from a covariant gauge-invariant formulation and used the gauge invariance to choose a light-cone gauge.[6] In the case of type II superstrings, as well as the field theory that describes the massless sector, there seems to be no manifestly covariant action principle, so it is necessary to use a physical-gauge formulation from the outset.[7] I will first describe the field theory (to order κ) and then indicate how the extension to the string theory works. Both cases use a "light-cone superspace" description involving eight anticommuting Grassmann coordinates θ^a that transform as a spinor under the SO(8) that rotates the eight transverse directions. In generalizing to the string, θ^a and the eight transverse coordinates x^i become functions of σ, a parameter that labels the points along the length of the string. Also, some of the super-Poincaré symmetries are realized locally in σ, while others are not. The locally conserved quantities correspond physically to continuity of $x^i(\sigma)$ and $\theta^a(\sigma)$ when strings interact, as well as local conservation of the conjugate momentum densities $p^i(\sigma)$ and $\lambda^a(\sigma)$.

2. Extended Supergravity in Ten Dimensions

There are two inequivalent extended supergravity theories in ten dimensions, both of which correspond to N=8 supergravity on truncation to four dimensions. The physical states of the two theories belong to different multiplets of SO(8), the group whose representations label massless states in ten dimensions, although they coincide on truncation to nine or fewer dimensions. One theory, containing two inequivalent gravitino representations, can be obtained by truncation of d=11 supergravity. The other, to be considered here, contains two equivalent gravitinos and cannot be obtained from a higher dimension. Besides the graviton and gravitinos it contains a pair of spinors, scalars, and second-rank antisymmetric tensors, as well as a single self-dual fourth-rank antisymmetric tensor. This is precisely the particle content described by a scalar superfield $\Phi(x,\theta)$, where θ^a is an 8-component spinor Grassman coordinate.

The spacetime manifold is described by light-cone coordinates $x^{\pm} = \frac{1}{\sqrt{2}}(x^0 \pm x^9)$ and transverse coordinates x^i, $i=1,2,\ldots,8$. The theory is conveniently described in a Hamiltonian formalism with H regarded as the generator of translations in x^+, the light-cone "time." The other coordinates are Fourier transformed so that

$$\alpha \equiv 2p^+ = 2i \frac{\partial}{\partial x^-}, \qquad (1)$$

$$p^i = -i \frac{\partial}{\partial x^i}, \qquad (2)$$

$$\lambda^a = \frac{\partial}{\partial \theta^a}, \qquad (3)$$

and we use a momentum-space field $\Phi(x^+, \alpha, p, \lambda)$. The expansion of this field in powers of λ contains 128 Bose modes and 128 Fermi modes and has its coefficients arranged so that its hermitian conjugate equals its Grassman Fourier transform

$$\Phi^\dagger(\alpha,p,\lambda) = \frac{\alpha^4}{16} \int \Phi(\alpha,p,\lambda') \, e^{2\lambda\lambda'/\alpha} d^8\lambda'. \qquad (4)$$

This property is crucial for demonstrating hermiticity of the interactions.[3]

The algebra of SO(8) involves eight matrices $\gamma^i_{a\dot{a}}$, where dotted and undotted spinor indices refer to inequivalent 8's of SO(8). These matrices satisfy the usual Dirac algebra. (An explicit representation is given in Appendix A of ref. 4). A supersymmetry charge in ten dimensions has 16 real components that decompose under transverse SO(8) into two eight-component spinors q^{+a} and $q^{-\dot{a}}$.

For the extended supergravity theory there are two of each, represented in terms of superspace coordinates and momenta as follows:

$$q_1^{+a} = \frac{\alpha}{\sqrt{2}}\,\theta^a \quad , \qquad q_1^{-\dot{a}} = (\gamma\cdot p\theta)^{\dot{a}} \quad , \tag{5a}$$

$$q_2^{+a} = \sqrt{2}\,\lambda^a \quad , \qquad q_2^{-\dot{a}} = \frac{2}{\alpha}\,(\gamma\cdot p\lambda)^{\dot{a}} \quad . \tag{5b}$$

These satisfy, in particular,

$$\{q_1^{-\dot{a}},\, q_2^{-\dot{b}}\} = 2h\delta^{\dot{a}\dot{b}} \quad , \tag{6}$$

where

$$h = p^2/\alpha \tag{7}$$

plays the role of the Hamiltonian, since the mass-shell condition is $h = p^-$. The Lorentz generators are represented in similar fashion:

$$j^{i+} = \frac{1}{2}\,x^i\alpha - x^+p^i \quad , \tag{8a}$$

$$j^{ij} = x^ip^j - x^jp^i - \frac{i}{2}\,\theta\gamma^{ij}\lambda \quad , \tag{8b}$$

$$j^{+-} = x^+h + i\alpha\,\frac{\partial}{\partial\alpha} - \frac{i}{2}\,\theta\lambda + \frac{5i}{2} \, , \tag{8c}$$

$$j^{i-} = x^i h + 2ip^i\,\frac{\partial}{\partial\alpha} - \frac{i}{\alpha}\,\theta\gamma^i\gamma\cdot p\lambda + \frac{3ip^i}{\alpha} \quad . \tag{8d}$$

When any generator is represented in terms of fields, it is denoted by the corresponding capital letter. In general for the free theory

$$G = \frac{1}{2}\int \alpha\Phi(-\alpha,\, -p,\, -\lambda)\,g\Phi(\alpha,p,\lambda)\,d\alpha d^8p d^8\lambda, \tag{9}$$

as is easily verified using the equal x^+ commutation relation

$$[\,\Phi(1),\Phi(2)\,] = \frac{1}{\alpha_1}\,\delta(\alpha_1 + \alpha_2)\delta^8(p_1 + p_2)\delta^8(\lambda_1+\lambda_2) \quad . \tag{10}$$

In the interacting field theory this form of G is exact for p^i, J^{i+}, J^{ij}, Q_1^{+a}, and Q_{2a}^+, but J^{+-}, J^{i-}, H, $Q_1^{-\dot{a}}$, and $Q_2^{-\dot{a}}$ include additional interaction terms involving cubic and higher powers of Φ. The Hamiltonian, in particular, has an expansion

$$H = H_2 + \kappa H_3 + \kappa^2 H_4 + \dots , \tag{11}$$

where H_2 has the form given in eq. (9).

The cubic contribution to H and the other generators, and undoubtedly the higher-order terms as well, are determined by the super-Poincaré algebra. The result is unique if one demands a correspondence with the massless sector of the string theory described in the next section. We find that

$$H_3 = \int d\mu_3^0 P^i P^j v^{ij}(\Lambda)\Phi(1)\Phi(2)\Phi(3) \tag{12}$$

$$Q_{cubic}^{-\dot{a}} = \frac{1}{\sqrt{2}} \left(\eta Q_1^{-\dot{a}} + \eta^* Q_2^{-\dot{a}} \right)_{cubic}$$

$$= \int d\mu_3^0 \, P^i s^{i\dot{a}} (\Lambda)\Phi(1)\Phi(2)\Phi(3) \tag{13a}$$

$$\tilde{Q}_{cubic}^{-\dot{a}} = \frac{1}{\sqrt{2}} \left(\eta^* Q_1^{-\dot{a}} + \eta Q_2^{-\dot{a}} \right)_{cubic}$$

$$= \int d\mu_3^0 P^i \tilde{s}^{i\dot{a}}(\Lambda)\Phi(1)\Phi(2)\Phi(3), \tag{13b}$$

where

$$\eta = e^{i\pi/4} \tag{14}$$

$$P^i = \alpha_1 p_2^i - \alpha_2 p_1^i \tag{15}$$

$$\Lambda^a = \alpha_1 \lambda_2^a - \alpha_2 \lambda_1^a \tag{16}$$

$$d\mu_3^0 = \left(\prod_{r=1}^{3} d\alpha_r d^8 p_r d^8 \lambda_r\right) \delta(\Sigma\alpha_r) \delta^8(\Sigma p_r) \delta^8(\Sigma\lambda_r) \ . \tag{17}$$

The quantities P^i and Λ^a effectively have total antisymmetry in the labels 1, 2, 3 when the δ functions in the measure are taken into account. Also, the functions ν^{ij}, $s^{i\dot{a}}$, and $\tilde{s}^{i\dot{a}}$ are polynomials in Λ uniquely determined by the normalization condition $\nu^{ij}(0) = \delta^{ij}$ and the equations

$$(\eta\Lambda^a + \frac{\alpha}{2}\eta^* \frac{\partial}{\partial\Lambda^a})\nu^{ij}(\Lambda) = \frac{i}{\sqrt{2}}\gamma^j_{a\dot{a}}s^{i\dot{a}}(\Lambda) \tag{18a}$$

$$(\eta^*\Lambda^a + \frac{\alpha}{2}\eta \frac{\partial}{\partial\Lambda^a})\nu^{ij}(\Lambda) = -\frac{i}{\sqrt{2}}\gamma^i_{a\dot{a}}\tilde{s}^{j\dot{a}}(\Lambda). \tag{18b}$$

These equations are derived by requiring that the anticommunication rules

$$\{Q^{-\dot{a}}, Q^{-\dot{b}}\} = \{\tilde{Q}^{-\dot{a}}, \tilde{Q}^{-\dot{b}}\} = 2H\delta^{\dot{a}\dot{b}} \tag{19}$$

$$\{Q^{-\dot{a}}, \tilde{Q}^{-\dot{b}}\} = 0 \tag{20}$$

are satisfied to order κ*. We expect the supersymmetry algebra to uniquely determine the higher-order terms as well, but the mathematics is somewhat complicated and has not yet been completed. The N = 8 theory in four dimensions is obtained by a trivial truncation, i.e., by dropping the dependence of the field on six of the eight transverse coordinates. The resulting emergence of an SU(8) symmetry is described in ref. [3].

3. Type II Superstrings

Type II superstrings are described by light-cone superspace coordinates $x^i(\sigma)$ and $\upsilon^a(\sigma)$ in position space or $p^i(\sigma)$ and $\lambda^a(\sigma)$ in momentum space, where σ is a parameter labeling points on the string in such a way that the density of p^+ is constant.

---

*Actually, this only gives ij symmetrical parts of eqs. (18), which by themselves do not have a unique solution. The ambiguity has been resolved by requiring a string generalization. A different choice was made in ref. [3].

Accordingly, the range of σ must be proportional to the total p^+ of the string, and is taken to be $-\pi|\alpha|\leq\sigma\leq\pi|\alpha|$. Since type II strings are closed, it is convenient to extend the range of σ, imposing the periodicity requirements

$$x^i(\sigma) = x^i(\sigma + 2\pi\alpha) \tag{21a}$$

$$\theta^a(\sigma) = \theta^a(\sigma + 2\pi\alpha), \tag{21b}$$

with similar relations for $p^i(\sigma)$ and $\lambda^a(\sigma)$. The field that creates or destroys a string is the scalar functional $\Phi[x^+,\alpha,p(\sigma),\lambda(\sigma)]$, subject to the constraint

$$\Phi\,[p(\sigma),\lambda(\sigma)] = \Phi[p(\sigma + \sigma_0),\lambda(\sigma + \sigma_0)]. \tag{22}$$

This corresponds to a rigid displacement of the parametrization along the string, which clearly should have no physical significance. It is the only reparametrization symmetry present in the light-cone formalism.

The superstring field theory can be written succintly in functional language. However, this conceals a number of subtle issues and may not be the most useful form for explicit calculations. In any case we have found it essential to insert explicit mode expansions and to study the interaction of specific string modes. In this way one always deals with unambiguous and well-defined expressions, which in the end can be reassembled into functional expressions. For this purpose the coordinates are expanded in Fourier series as follows (in units with $\alpha' = 1$):

$$x^i(\sigma) = x^i + \sum_{n\neq 0} \frac{1}{n}\left[\alpha_n^i e^{in\sigma/|\alpha|} + \tilde{\alpha}_n^i\, e^{-in\sigma/|\alpha|}\right] \tag{23}$$

$$p^i(\sigma) = \frac{1}{2\pi|\alpha|}\,[p^i + \frac{1}{2}\,\Sigma_{n\neq 0}\,(\alpha_n^i e^{in\sigma/|\alpha|} + \tilde{\alpha}_n^i e^{-in\sigma/|\alpha|})] \qquad (24)$$

$$q^{+a}(\sigma) = \eta^*\lambda^a(\sigma) + \frac{e(\alpha)}{4\pi}\,\eta\theta^a(\sigma) = \frac{1}{2\pi|\alpha|}\sum_{-\infty}^{\infty} Q_n^a\, e^{in\sigma/|\alpha|} \qquad (25a)$$

$$\tilde{q}^{+a}(\sigma) = \eta\lambda^a(\sigma) + \frac{e(\alpha)}{4\pi}\,\eta^*\theta^a(\sigma) = \frac{1}{2\pi|\alpha|}\sum_{-\infty}^{\infty} \tilde{Q}_n^a e^{in\sigma/|\alpha|}, \qquad (25b)$$

where

$$e(\alpha) \equiv \frac{\alpha}{|\alpha|}, \qquad (26)$$

and the oscillators have the commutation relations

$$[\alpha_m^i, \alpha_n^j] = [\tilde{\alpha}_m^i, \tilde{\alpha}_n^j] = m\delta^{ij}\delta_{m+n,0} \qquad (27)$$

$$\{Q_m^a, Q_n^b\} = \{\tilde{Q}_m^a, \tilde{Q}_n^b\} = \alpha\delta^{ab}\delta_{m+n,0} \qquad (28)$$

$$[\alpha_m^i, \tilde{\alpha}_n^j] = \{Q_m^a, \tilde{Q}_n^b\} = 0, \qquad (29)$$

so that

$$[x^i(\sigma), p^j(\sigma')] = i\delta^{ij}\delta(\sigma - \sigma') \qquad (30)$$

$$\{\theta^a(\sigma), \lambda^b(\sigma')\} = \delta^{ab}\delta(\sigma - \sigma') \qquad (31)$$

$$[x^i(\sigma), x^j(\sigma')] = [p^i(\sigma), p^j(\sigma')] = 0 \qquad (32)$$

$$\{\theta^a(\sigma), \theta^b(\sigma')\} = \{\lambda^a(\sigma), \lambda^b(\sigma')\} = 0 \qquad (33)$$

Note that positive subscripts represent lowering operators and negative subscripts give the corresponding raising operators.

The Hamiltonian generalizing eq. (7) is given by

$$h = \int_{-\pi|\alpha|}^{\pi|\alpha|} \left[2\pi e(\alpha) p^2 + \frac{e(\alpha)}{8\pi} [x'(\sigma)]^2 + 2\pi\lambda'(\sigma)\lambda(\sigma) - \frac{1}{8\pi} \theta'(\sigma) \times \theta(\sigma)\right] d\sigma = \frac{1}{\alpha}(p^2 + N + \tilde{N}) \tag{34}$$

where

$$N = \sum_{n=1}^{\infty} (\alpha^i_{-n}\alpha^i_n + \frac{n}{\alpha} Q^a_{-n} Q^a_n) \tag{35a}$$

$$\tilde{N} = \sum_{n=1}^{\infty} (\tilde{\alpha}^i_{-n}\tilde{\alpha}^i_n + \frac{n}{\alpha} \tilde{Q}^a_{-n} \tilde{Q}^a_n). \tag{35b}$$

In terms of these operators, the constraint in eq.(22) takes the form $(N - \tilde{N}) = 0$. Normal ordering the oscillators in the integrand of eq. (34) gives (infinite) constants that cancel between the bosonic and fermionic terms, so the expression is correct as written. This feature is one of many advantages of superstrings over older nonsupersymmetrical string models.

The supersymmetries in eq. (25) are locally conserved (as functions of σ) in string interactions. This means that $e(\alpha)\theta^a(\sigma)$ and the Grassmann momentum $\lambda^a(\sigma)$ are conserved. The former can be interpreted as expressing continuity of the coordinate between initial and final string configurations. ($e(\alpha)$ is positive for incoming strings and negative for outgoing ones.) In similar fashion $e(\alpha)x^i(\sigma)$ and the ordinary momentum density $p^i(\sigma)$ are also conserved locally. These may also be interpreted as consequences of local conservation of j^{i+} density

$$j^{i+}(\sigma) = \frac{1}{\pi} e(\alpha) x^i(\sigma) - x^+ p^i(\sigma). \tag{36}$$

The q^- generators, on the other hand, describe global symmetries only. The formulas are

$$q_1^- = \int_{-\pi|\alpha|}^{\pi|\alpha|} [\gamma\cdot p(\sigma)\theta(\sigma) + ie(\alpha)\gamma\cdot x'(\sigma)\lambda(\sigma)]d\sigma \tag{37a}$$

$$q_2^- = \int_{-\pi|\alpha|}^{\pi|\alpha|} [4\pi e(\alpha)\gamma\cdot p(\sigma)\lambda(\sigma) - \frac{i}{4\pi}\gamma\cdot x'(\sigma)\theta(\sigma)]d\sigma \tag{37b}$$

or

$$q^- = \frac{1}{\sqrt{2}}(\eta q_1^- + \eta^* q_2^-) = \frac{\sqrt{2}}{\alpha}\sum_{-\infty}^{\infty} \gamma\cdot\alpha_{-n}Q_n \tag{38a}$$

$$\tilde{q}^- = \frac{1}{\sqrt{2}}(\eta^* q_1^- + \eta q_2^-) = \frac{\sqrt{2}}{\alpha}\sum_{-\infty}^{\infty} \gamma\cdot\tilde{\alpha}_{-n}\tilde{Q}_n . \tag{38b}$$

The passage to representations of the algebra in terms of fields works essentially the same as in the field theory with

$$G = \frac{1}{2}\int \alpha\Phi[-\alpha,-p(\sigma),-\lambda(\sigma)]g\ \Phi[\alpha,p(\sigma),\lambda(\sigma)]d\alpha D^8 p(\sigma) D^8\lambda(\sigma) \tag{39}$$

replacing eq. (9) and

$$[\Phi(1),\Phi(2)] = \frac{1}{\alpha_1}\delta(\alpha_1 + \alpha_2)\Delta^8[p_1(\sigma) + p_2(\sigma)]\Delta^8[\lambda_1(\sigma) + \lambda_2(\sigma)] \tag{40}$$

replacing eq. (10). The functional Δ functions and functional integrations are given a precise meaning as infinite products over the sine and cosine modes of the Fourier expansions.

Consider now the three-string interaction shown in fig. 1, in which incoming strings #1 and #2 join to form outgoing string #3. A parameter σ_r is associated with string r according to the rule

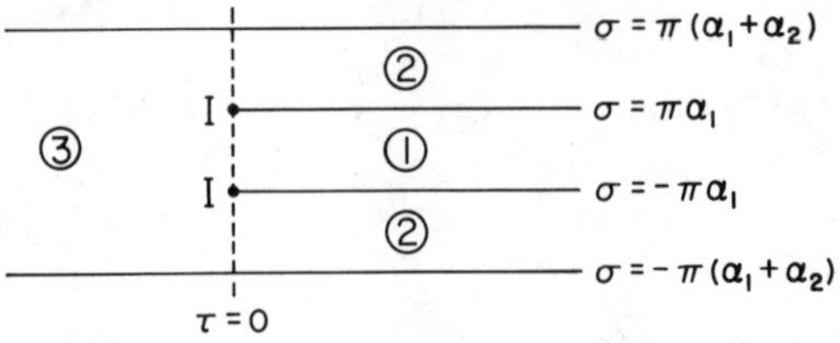

Fig. 1. Strings #1 and #2 joint at I to form string #3. The boundaries of regions 1,2, and 3 are identified so as to describe closed strings.

$$\sigma_1 = \sigma \qquad -\pi\sigma_1 \leq \sigma \leq \pi\alpha_1 \tag{41a}$$

$$\sigma_2 = \begin{cases} \sigma - \pi\alpha_1 & \pi\alpha_1 \leq \sigma \leq \pi(\alpha_1 + \alpha_2) \\ \sigma + \pi\alpha_1 & -\pi(\alpha_1 + \alpha_2) \leq \sigma \leq \pi(\alpha_1 + \alpha_2) \end{cases}$$

$$\sigma_3 = -\sigma \qquad -\pi(\alpha_1 + \alpha_2) \leq \sigma \leq \pi(\alpha_1 + \alpha_2). \tag{41c}$$

Then the momentum densities are given by

$$p_r^i(\sigma) = \Theta_r p_r^i(\sigma_r) \quad , \quad \lambda_r^a(\sigma) = \Theta_r \lambda_r^a(\sigma_r) \quad , \tag{42,43}$$

where the Θ_r's are the step functions

$$\Theta_1 = \theta(\pi\alpha_1 - |\sigma|) \ , \ \Theta_2 = \theta(|\sigma| - \pi_{\alpha 1}) \ , \ \Theta_3 = \Theta_1 + \Theta_2 = 1 \tag{44a,b,c}$$

With these definitions and analogous ones for $x_r(\sigma)$ and $\theta_r(\sigma)$ the local conservation laws are

$$\sum_{r=1}^{3} p_r^i(\sigma) = \sum_{r=1}^{3} \lambda_r^a(\sigma) = 0 \tag{45}$$

$$\sum_{r=1}^{3} e(\alpha_r) x_r^i(\sigma) = \sum_{r=1}^{3} e(\alpha_r) \theta_r^a(\sigma) = 0. \tag{46}$$

The cubic interaction Hamiltonian is written in the form

$$H_3 = \int d\mu_3 G\Phi(1)\Phi(2)\Phi(3), \tag{47}$$

where

$$d\mu_3 = \left[\prod_{r=1}^{3} d\alpha_r D^8\lambda_r(\sigma) D^8 p_r(\sigma)\right] \delta(\Sigma\alpha_r)\Delta^8[\Sigma p_r(\sigma)]\Delta^8[\Sigma\lambda_r(\sigma)], \tag{48}$$

and G is an operator to be determined. The local conservation laws in eqs. (45) and (46) are satisfied as a consequence of the Δ functionals in eq. (48) provided that G is of suitable form. The string fields Φ may be regarded as infinite component fields containing an ordinary point-particle field for every set $\{n\}$ of excitations of the various Bose and Fermi oscillators (subject to the constraint of eq. (22)). Therefore the information contained in eq. (47) is equivalent to an infinite set of couplings $C(\{n^{(1)},n^{(2)},n^{(3)}\})$ describing the interaction of three arbitrary string states. This information is conveniently collected in a giant ket vector

$$|V\rangle = \sum_{\{n^{(1)},n^{(2)},n^{(3)}\}} C(\{n^{(1)},n^{(2)},n^{(3)}\})|\{n^{(1)},n^{(2)},n^{(3)}\}\rangle, \tag{49}$$

where we have associated a Fock-space state with each set of mode numbers. This form of the vertex can be calculated by inserting explicit mode expansions for the fields in eq. (47) and doing the integrals, all of which are Gaussians. The result has the form

$$|V\rangle = G_{op}E_\alpha E_Q|0\rangle\delta(\Sigma\alpha_r)\delta^8(\Sigma p_r)\delta^8(\Sigma\lambda_r), \tag{50}$$

where G_{op} corresponds to G in eq. (47), and

$$E_\alpha = \exp\left\{ \frac{1}{2} \sum (\alpha_{-m}^{(r)} \bar{N}_{mn}^{rs} \alpha_{-n}^{(s)} + \tilde{\alpha}_{-m}^{r} \bar{N}_{mn}^{rs} \tilde{\alpha}_{-n}^{s}) + P \sum \bar{N}_m^r (\alpha_{-m}^{(r)} + \tilde{\alpha}_{-m}^{(r)}) - \frac{\tau_0}{\alpha} P^2 \right\} \tag{51}$$

$$E_Q = \exp\left\{ \frac{1}{2} \sum \frac{m}{\alpha_r} (Q_{-m}^{(r)} \bar{N}_{mn}^{rs} Q_{-n}^{(s)} + \tilde{Q}_{-m}^{(r)} \bar{N}_{mn}^{rs} \tilde{Q}_{-n}^{(s)}) + \frac{i}{2}\alpha\left(\sum Q_{-m}^{(r)} \frac{m}{\alpha_r} \bar{N}_m^r\right)\left(\sum \tilde{Q}_{-n}^{(s)} \frac{n}{\alpha_s} \bar{N}_m^s\right) - \Lambda \sum \frac{m}{\alpha_r} \bar{N}_m^r \left(\eta^* Q_{-m}^{(r)} + \eta \tilde{Q}_{-m}^{(r)}\right) \right\}. \tag{52}$$

In these expressions

$$\alpha = \alpha_1 \alpha_2 \alpha_3 \tag{53}$$

$$\tau_0 = \sum_{r=1}^{3} \alpha_r \ln|\alpha_r| \tag{54}$$

$$\bar{N}_{mn}^{rs} = -\frac{mn\alpha}{n\alpha_r + m\alpha_s} \bar{N}_m^r \bar{N}_n^s \tag{55}$$

$$\bar{N}_m^r = \frac{1}{\alpha_r} f_m\left(-\frac{\alpha_{r+1}}{\alpha_r}\right) e^{m\tau_0/\alpha_r} \tag{56}$$

$$f_m(\gamma) = \frac{1}{m!} \frac{\Gamma(m\gamma)}{\Gamma(m\gamma + 1 - m)} \tag{57}$$

and P and Γ are as given in eqs. (15) and (16). Using various identities among these quantities, one can show that the sums in eqs. (45) and (46) annihilate $|V\rangle$ if they commute with G_{op}. The next step is to determine G_{op} in eq. (50). The first requirement is that it should commute with the quantities in eqs. (45) and (46). This is satisfied by the three linear expressions

$$X = P - \sum_{r=1}^{3} \sum_{n=1}^{\infty} \frac{n\alpha}{\alpha_r} \bar{N}_n^r \alpha_{-n}^{(r)}, \quad \tilde{X} = P - \sum_{r=1}^{3} \sum_{n=1}^{\infty} \frac{n\alpha}{\alpha_r} \bar{N}_n^r \tilde{\alpha}_{-n}^{(r)} \tag{58a,b}$$

$$Y = \Lambda - \frac{1}{2} \sum_{r=1}^{3} \sum_{n=1}^{\infty} \frac{n\alpha}{\alpha_r} \bar{N}^r_n \left[Q^{(r)}_{-n} + \tilde{Q}^{(r)}_{-n} \right] . \tag{59}$$

There is one slightly subtle case, namely

$$[X - \tilde{X}, \sum p_r(\sigma)] \propto \delta(\sigma - \pi\alpha_1) - \delta(\sigma + \pi\alpha_1) , \tag{60}$$

which appears to violate one of the conditions. However, $\sigma = \pi\alpha_1$ and $\sigma = -\pi\alpha_1$ correspond to the same spacetime point, i.e., the place where the interaction occurs, and therefore the total momentum density at the interaction point is conserved. We now rewrite eq. (50) and the cubic supersymmetry operators in the form

$$|V\rangle = \tilde{X}^i X^j \nu^{ij}(Y) E_\alpha E_Q |0\rangle \, \delta^{17} \tag{61}$$

$$|Q^{-\dot{a}}\rangle = \tilde{X}^i s^{i\dot{a}}(Y) E_\alpha E_Q |0\rangle \, \delta^{17} \tag{62a}$$

$$|\tilde{Q}^{-\dot{a}}\rangle = X^i \tilde{s}^{i\dot{a}}(Y) E_\alpha E_Q |0\rangle \, \delta^{17} , \tag{62b}$$

which automatically builds in the local symmetry requirements and reduces the problem to the determination of functions $\nu^{ij}, \tilde{s}^{i\dot{a}}$, and $s^{i\dot{a}}$. Implementing the supersymmetry algebra of eq. (19) to order κ leads to eqs. (18a and b) with Λ replaced by Y. Equation (20) in order κ then serves as a nontrivial consistency check on the solution. Explicit expansions of ν, $\tilde{s}$, and s are given in Appendix D of ref. [5].

The functional equivalents of these oscillator results can now be described. The basic idea is that there are derivative couplings at the interaction point $\sigma = \pm\pi\alpha_1$. However, these are singular and need to be described as limits. Thus, we find by explicit comparison with the oscillator expressions that

$$H_3 = \lim_{\sigma \to \pi\alpha_1} \int d\mu_3 \tilde{X}^i(\sigma) \, X^j(\sigma) \nu^{ij}(Y(\sigma)) \Phi(1)\Phi(2)\Phi(3) \tag{63}$$

$$Q^{-\dot{a}}_{cubic} = \lim_{\sigma\to\pi\alpha_1} \int d\mu_3 \tilde{X}^i(\sigma) s^{i\dot{a}}(Y(\sigma))\Phi(1)\Phi(2)\Phi(3) \tag{64a}$$

$$\tilde{Q}^{-\dot{a}}_{cubic} = \lim_{\sigma\to\pi\alpha_1} \int d\mu_3 X^i(\sigma) \tilde{s}^{i\dot{a}}(Y(\sigma))\Phi(1)\Phi(2)\Phi(3), \tag{64b}$$

where

$$X(\sigma) = -2\pi\sqrt{-\alpha}(\pi\alpha_1-\sigma)^{1/2}[p^{(1)}(\sigma) - \frac{1}{4\pi}x^{(1)'}(\sigma) + p^{(1)}(-\sigma) - \frac{1}{4\pi}x^{(1)'}(-\sigma)] \tag{65a}$$

$$\tilde{X}(\sigma) = -2\pi\sqrt{-\alpha}(\pi\alpha_1-\sigma)^{1/2}[p^{(1)}(\sigma) + \frac{1}{4\pi}x^{(1)'}(\sigma) + p^{(1)}(-\sigma) + \frac{1}{4\pi}x^{(1)'}(-\sigma)] \tag{65b}$$

$$Y(\sigma) = -2\pi\sqrt{-\alpha}(\pi\alpha_1-\sigma)^{1/2}[\lambda^{(1)}(\sigma) + \lambda^{(1)}(-\sigma)]. \tag{66}$$

The vanishing factors in these expressions combine with the singular behavior of the operations to give the desired result in the limit. Equations (65) and (66) are expressed in terms of the variables of string #1, but in the limit the choice of any strings is equivalent, and the vertex, in fact, has total symmetry in the three strings.

It is important to complete the construction of the field theory of sect. 2 and the string theory of this section. Both are expected to be nonpolynomial. Once this is achieved, we should be in a position to look for classical solutions that describe physically interesting compactifications and to study the quantum behavior of the theories more completely than has been possible with previous formalisms. The prospects are quite exciting.

REFERENCES

1. J. Scherk and J.H. Schwarz, Nucl. Phys. B81 (1974) 118; Phys. Lett. 57B (1975) 463; J.H. Schwarz in "New Frontiers in High-Energy Physics," Proc. Orbis Scientiae 1978, ed. A. Perlmutter

and L.F. Scott (Plenum, New York, 1978) p. 431.
2. J.H. Schwarz "Superstring Theory," Phys. Reports Vol. 89, No. 3, 1982 and references therein.
3. M.B. Green and J.H. Schwarz, Caltech preprint CALT-68-957, to be published in Phys. Lett. B.
4. M.B. Green and J.H. Schwarz, Caltech preprint CALT-68-956, to be published in Nucl. Phys. B.
5. M.B. Green, J.H. Schwarz, and L. Brink, Caltech preprint CALT-68-972.
6. P. Goddard, J. Goldstone, C. Rebbi, and C.B. Thorn, Nucl. Phys. B56 (1973) 109.
7. N. Marcus and J.H. Schwarz, Phys. Lett. 115B (1982) 111.

GRAVITATIONAL GAUGE FIELDS*

Heinz R. Pagels

The Rockefeller University

New York, New York 10021

In spite of the experimental success of Einstein's classical theory of gravity - general relativity - the theory leaves much to be desired if we examine it from the viewpoint of fundamental quantum field theory. One problem is that the theory is not renormalizable if coupled to matter. Another problem is that the Euclidean action is not bounded from below for arbitrary metrics and consequently, the functional path integral is not well defined. Furthermore, there is the puzzle of why the cosmological term is zero or near zero in our universe - a fact that deserves a fundamental explanation.

Perhaps the most intriguing problem of all is that of unifying gravity with the other interactions, a problem which is compounded by the realization that all the other interactions are mediated by spin one gauge fields while gravity is a spin two field. Is it possible that this distinction is superficial and that gravity can also be viewed as an ordinary gauge field theory? The point of this talk is to give indications that the answer to this question is yes.

Beginning with the work of Utiyama[1] (1956) and Kibble[2] (1961), the analogy between ordinary Yang-Mills gauge fields and gravity has

---

*Work supported in part by the U.S. Department of Energy under Contract Grant No. DE-AC02-83ER40033.B000.

been stressed. One aspect of this development - an observation of McDowell and Mansouri[3] - is of special interest for the model we will subsequently describe.

Consider an O(5) gauge field $A_\mu^{ij} = -A_\mu^{ji}$, i, j = 1,2,3,4,5 and the field strength

$$F_{\mu\nu}^{ij}(A) = \partial_\mu A_\nu^{ij} - \partial_\nu A_\mu^{ij} + A_\mu^{ik} A_\nu^{kj} - A_\nu^{ik} A_\mu^{kj} \quad ,$$

defined on a base manifold with Euclidean signature $g_{\mu\nu} = [++++]$. MacDowell and Mansouri consider the following action

$$S = \int d^4x \, F_{\mu\nu}^{ij}(A) \, F_{\lambda\delta}^{\ell m}(A) \, \varepsilon_{ij\ell m} \, \varepsilon^{\mu\nu\lambda\delta} \quad .$$

Because the tensor $\varepsilon^{\mu\nu\lambda\delta}$ is a density, the action S is independent of the metric $g_{\mu\nu}$ of the base manifold. Note the way the O(5) indices are contracted with the O(4) tensor $\varepsilon_{ij\ell m}$. S is not O(5) invariant (it is O(4) invariant). In particular, it is not a Pontryagin index although it "almost" is. The physical content of this action becomes more transparent if we identify the 10 O(5) gauge fields in terms of 6 fields in the adjoint representation of O(4) and 4 fields in the vector representation (we single out the "5" direction)

$$A_\mu^{i5} = \varepsilon_\mu^i \quad , \quad i = 1 \ldots 4,$$

$$A_\mu^{ij} = \omega_\mu^{ij} \quad , \quad i, j = 1 \ldots 4 \; .$$

Then

$$F_{\mu\nu}^{ij}(A) = R_{\mu\nu}^{ij}(\omega) + \varepsilon_\mu^i \varepsilon_\nu^j - \varepsilon_\nu^i \varepsilon_\mu^j \quad , \; i, j = 1 \ldots 4,$$

$$R_{\mu\nu}^{ij}(\omega) = \partial_\mu \omega_\nu^{ij} - \partial_\nu \omega_\mu^{ij} + \omega_\mu^{ik} \omega_\nu^{kj} - \omega_\nu^{ik} \omega_\mu^{kj} \quad ,$$

where $R^{ij}_{\mu\nu}(\omega)$ is the O(4) field strength. Using

$$F^{ij}_{\mu\nu}(A)F^{\ell m}_{\lambda\delta}(A)\varepsilon_{ij\ell m}\varepsilon^{\mu\nu\lambda\delta} = R^{ij}_{\mu\nu}(\omega)R^{\ell m}_{\lambda\delta}(\omega)\varepsilon^{\mu\nu\lambda\delta}\varepsilon_{ij\ell m}$$

$$+ 4R^{ij}_{\mu\nu}(\omega)\varepsilon^{\ell}_{\lambda}\varepsilon^{m}_{\delta}\varepsilon_{ij\ell m}\varepsilon_{ij\ m}\varepsilon^{\mu\nu\lambda\delta} + 44!\ \det \varepsilon^{i}_{\mu} \quad ,$$

we find the action becomes

$$S = 4\int d^4x(R^{ij}_{\mu\nu}(\omega)\varepsilon^{\ell}_{\lambda}\varepsilon^{m}_{\delta}\varepsilon_{ij\ell m}\varepsilon^{\mu\nu\lambda\delta} + 4!\ \det \varepsilon^{i}_{\mu} \quad ,$$

upon dropping the O(4) Pontryagin index which doesn't contribute to equations of motion. Since this action is independent of $g_{\mu\nu}$ we are free to define the metric as $g_{\mu\nu} = \varepsilon^{a}_{\mu}\varepsilon^{a}_{\nu}$. Then the above action is precisely the Hilbert action with a cosmological term in the first order formalism. If we vary the action with respect to ω_{μ}^{ab} the extremal principle yields the usual relation between the connection and vierbein. The odd action we began with is really rather familiar.

The problem with the MacDowell-Mansouri action is that no one could find a reasonable way to couple it to matter. However, if one instead asks the question: "Can the MacDowell-Mansouri action be obtained as the effective action of an O(5) gauge theory?", the answer is yes and the matter coupling problem takes care of itself.

Consider the O(5) gauge theory, with fermions ψ^{ij} in the adjoint representation and a real pseudoscalar field ϕ^{i} in the vector representation.[4]

The action on a base manifold with vierbein $e^{a}{}_{\mu}$ is

$$S = \int d^4x e L \quad , \qquad e = \det e^{a}_{\mu} \quad .$$

$$L = L_g + L_F + L_S + L_Y + L_V + L_R + L_R 2$$

contains all dimension 4 operators consistent with even parity and gauge invariance which are

$$\mathcal{L}_g = \frac{1}{4g^2} F^{ij}_{\mu\nu} F^{ij}_{\lambda\delta} g^{\lambda\mu} g^{\nu\delta} \quad , \quad g_{\mu\nu} = e^a_\mu e^a_\nu \quad ,$$

$$\mathcal{L}_F = i e^a_\mu \bar{\psi}^{ij} \gamma_a D_\mu \psi^{ij} \quad ,$$

$$\mathcal{L}_S = \frac{1}{2} D_\mu \phi^i D_\nu \phi^i g^{\mu\nu} + \lambda(\phi^{i^2} - \mu^2)^2 \quad ,$$

$$\mathcal{L}_Y = g_F \bar{\psi}^{ij} i\gamma_5 \psi^{\ell m} \phi^k \varepsilon_{ij\ell mk} \quad ,$$

$$\mathcal{L}_V = \Lambda^{14} \quad ,$$

$$\mathcal{L}_R = (\varepsilon_1 \mu^2 + \varepsilon_2 \phi^{i^2}) \; R(g) \quad ,$$

$$\mathcal{L}_{R^2} = \gamma_1 R_{\mu\nu\lambda\delta} R^{\mu\nu\lambda\delta} + \gamma_2 R_{\mu\nu} R^{\mu\nu} + \gamma_3 R^2 \quad .$$

Here the covariant derivatives include both the geometrical connection and the gauge field affinity. We will ignore the term $\mathcal{L}_R$ in what follows for the interest of clarity. The R^2 terms are small near flat space and do not effect the usual solar tests of relativity. If we ignore these terms, what remains is just a gauge theory on a Riemannian manifold.

Because of the choice of the scalar field potential, the O(5) symmetry breaks with a scale of order μ which we assume to be near the Planck mass. A gauge choice lets us write for the only non-vanishing component of the ϕ^i field $\phi^5 = \mu + \phi$ where the physical field ϕ has a mass on the order of the Planck mass. Likewise, the fermions (except for ψ^{a5}) get masses on the order of the Planck mass.

We are interested in constructing the low energy (sub-Planck mass) effective action in the bosonic sector. So far, the only dimension 4 operators in this effective action are

$$\mathcal{L}_4^{\text{eff}} = \frac{1}{4g^2} F^{ij}_{\mu\nu} F^{ij}_{\lambda\delta} g^{\mu\lambda} g^{\nu\delta} + \frac{1}{2} \mu^2 \varepsilon^i_\mu \varepsilon^i_\nu g^{\mu\nu} + \Lambda^{14} \quad ,$$

where the F^2 term can be decomposed according to

$$F^{ij}_{\mu\nu}F^{ij}_{\lambda\delta}g^{\mu\lambda}g^{\nu\delta} = [2(D_\mu\varepsilon^i_\nu - D_\nu\varepsilon^i_\mu)(D_\lambda\varepsilon^i_\delta - D_\delta\varepsilon^i_\lambda)$$

$$+ R^{ij}_{\mu\nu}(\omega)R^{ij}_{\lambda\delta}(\omega) - 4\, R^{ij}_{\mu\nu}(\omega)\varepsilon^i_\lambda\varepsilon^j_\delta - 4\varepsilon^i_\mu\varepsilon^j_\nu\varepsilon^i_\lambda\varepsilon^j_\delta]g^{\mu\nu}g^{\lambda\delta}$$

in terms of O(4) fields. If we examine dimension 5 operators, we find that there are four independent ones on mass shell and only one of these contributes to the effective bosonic action.[4] This is a fermion triangle loop operator (the only reason for putting fermions into the model was to generate this operator) and is given by

$$0_1 = F^{ij}_{\mu\nu}F^{\ell m}_{\lambda\delta}\phi^k\varepsilon^{\mu\nu\lambda\delta}\varepsilon_{ij\ell mk} \quad .$$

In the broken symmetry phase it contributes to the action a term

$$e\mathcal{L}^{eff}_5 = G_\mu F^{ij}_{\mu\nu}F^{\ell m}_{\lambda\delta}\varepsilon_{ij\ell m}\varepsilon^{\mu\nu\lambda\delta}$$

where

$$G = \frac{g_F C}{\mu 16\pi^2} \quad ,$$

and C is proportional to the Casimir invariant for the fermions. This term is, of course, just the MacDowell-Mansouri action. Except for the F^2 term in $\mathcal{L}_4^{eff}$ we have accomplished our goal of getting this action from a gauge theory. Can one get rid of the F^2 term?

If the gauge theory is asymptotically free then the effective coupling g^2 is very large, even infinite, for low energy physics. The dimension five operator is independent of g^2 so it remains in the strong coupling regime but the $g^{-2}F^2$ term is driven to zero. At what mass scale can we expect this to occur? Suppose we imbed O(5) in O(6)~SU(4) and imagine a totally unified theory (TUT) which is SU(9). The SU(9) theory breaks as follows

$$SU(9)\to SU(4)\times SU(5)$$

$$M_p$$

$$SU(5)\to SU(3)\times SU(2)\times U(1)$$

$$M_x$$

where M_p is the Planck mass 10^{19}GeV and $M_x \sim 10^{15}$GeV is the GUT mass. The color SU(3) interactions with coupling constant become strong at $M_{QCD}=M_3\sim 1$GeV and we now ask at what mass scale M_4 do the SU(4) interactions become as strong so that $\alpha_4(M_4)=\alpha_3(M_3)$? The answer is[5]

$$M_4^4 = M_3^3 M_x^2/M_p \quad ,$$

so that $M_4 \sim 500$GeV, a number which is safely above the mass scale of presently observed particles. We conclude on this basis that the $g^{-2}F^2$ term is negligible in the low energy domain and that the dimension 5 operator dominates.

If we adopt this approach, then there is a problem in interpretation. Strong coupling is associated with confinement and all 0(4) nonsinglet fields like $\varepsilon^i_{\ \mu}$ and $\omega^{ij}_{\ \ \mu}$ should confine. Since we want to identify these fields with the vierbein and connection respectively, confinement does not seem desirable in this instance. However, we should bear in mind that singlet combinations of these fields like $\varepsilon^a_{\ \mu}\varepsilon^a_{\ \nu}$ and $R^{ab}_{\ \ \nu\mu}(\omega)\varepsilon^a_{\ \lambda}\varepsilon^b_{\ \delta}$ are observable and these will correspond to geometric objects like the metric and the Riemann curvature. So the interpretative problem may not be a difficulty.

An alternative approach is to destabilize the model in the ultraviolet region by adding more fermions and scalars so that it is not asymptotically free. Then the fields $\varepsilon^a_{\ \mu}$ and $\omega^{ab}_{\ \ \mu}$ are not confined and can be directly identified with the verbein and connection. The $g^{-2}F^2$ term can be rendered harmless by assuming that g^2 is a very large (but finite) number in the infrared region. While free of interpretative problems, this approach seems less desirable from the point of view of total unification which, in its simplest version, requires all fields to beome components of a single non-Abelian asymptotically free gauge field. Yet, this might be the right way to look at our model. In what follows, we assume we get rid of the F^2 term because of strong coupling.

The effective action is then

$$eL^{eff} = e(\frac{1}{2}\mu^2 \varepsilon^i_\mu \varepsilon^i_\nu g^{\mu\nu} - \Lambda^4) + G_\mu F^{ij}_{\mu\nu} F^{\ell m}_{\lambda\delta} \varepsilon_{ij\ell m} \varepsilon^{\mu\nu\lambda\delta}$$

(we have assumed the induced vacuum energy is negative). Extremizing $S^{eff} = \int eL^{eff} d^4x$ with respect to $g_{\mu\nu}$ we obtain

$$g_{\mu\nu} = \frac{\mu^2}{\Lambda^4} \varepsilon^a_\mu \varepsilon^a_\nu \quad ,$$

so the vierbein can be identified as $e^a_\mu = \frac{\mu}{\Lambda^2} \varepsilon^a_\mu$. Substituting this result back into the action, we find

$$S^{eff} = \int d^4x(- \frac{1}{8k^2} R^{ij}_{\mu\nu}(\omega)\ e^\ell_\lambda e^m_\delta \varepsilon_{ij\ell m} \varepsilon^{\mu\nu\lambda\delta} + \bar{\Lambda}^4 e) ,$$

with

$$k^2 = - \frac{\mu^2 \pi}{2g_F C\Lambda^2} \quad , \quad \frac{\bar{\Lambda}^4}{\Lambda^4} = 1 - \frac{3}{k^2 \mu^2} \quad ,$$

which is just the Hilbert action with a cosmological term. Setting the latter to zero, $\bar{\Lambda}=0$ implies $\kappa^2=3/\mu^2>0$, corresponding to attractive gravity.

It is not difficult to add matter to this effective action (assuming the matter consists of O(5) singlets). The presence of matter is formally implemented by adding to L^{eff} the term L^m which responds to a metric variation according to $\delta(\sqrt{g}L^m)=\sqrt{g}\theta_{\mu\nu}\delta g^{\mu\nu}$, where $\theta_{\mu\nu}$ is the energy momentum tensor. This modifies the previous equation for the metric-gauge field relation according to

$$g_{\mu\nu} = \frac{\mu^2}{\Lambda^4} \varepsilon^i_\mu \varepsilon^i_\mu + \frac{1}{\Lambda^4} (2\theta_{\mu\nu} - g_{\mu\nu}\theta) \quad .$$

If Λ and μ are on the order of the Planck mass, and since the scale of $\theta_{\mu\nu}$ is assumed to be less than this, $(\theta_{\mu\nu}/\Lambda^4)<<0$, we find

$$\varepsilon^a_\mu = \frac{\Lambda^2}{\mu} e^a_\mu (1 + \frac{\theta}{2\Lambda^4}) - \frac{\theta^i_\mu}{\Lambda^2 \mu} + 0(\frac{\theta_{\mu\nu}}{4})^2$$

$$\det \varepsilon^a_\mu = (\frac{\Lambda^4}{\mu^2})^2 (1 + \frac{\theta}{\Lambda^4}) \det e^a_\mu + 0(\frac{\theta_{\mu\nu}}{\Lambda^4})^2 \quad ,$$

$$\theta^i_\mu = \varepsilon^i_\lambda \theta^\lambda_\mu \quad .$$

Subsituting this back into the effective action, keeping only leading order terms in $\theta_{\mu\nu}/\Lambda^4$ and setting $\kappa^2=3/\mu^2$ so the cosmological terms vanish, we obtain

$$S^{eff} = \int d^4x(- \frac{1}{8k^2} R^{ij}_{\mu\nu}(\omega) e^\ell_\lambda e^m_\delta \varepsilon_{ij\ell m} \varepsilon^{\mu\nu\lambda\delta} + eL^m),$$

the conventional action of gravity coupled to matter. This is our main result.

The interesting feature of this model of gravity is that the metric is not the fundamental object. The dynamical field is the O(5) gauge field and the metric just goes along for the ride. The metric tensor is a quadratic form given by

$$g_{\mu\nu} = \sum \varepsilon^a_\mu \varepsilon^a_\nu \quad ,$$

where the sum is on the gauge field components of G/O(4) with $G \supseteq O(5)$. In this way, gravity can be unified with other gauge field models.

Yet, there are problems with this model. The first problem is that we ignored the R term in the action. If we include it in our list of dimension 4 operators, then the constraint equation for the metric becomes a dynamical equation. Furthermore, the gravity constant is no longer related to the mass scale μ. By including R term, we have explicitly included gravity from the very beginning.

S. Adler[6] in his review of induced gravity states the conditions under which the R term is not present. One possibility is to construct supersymmetric models which eliminate the R term naturally. Another possibility is to eliminate the scalar fields and appeal to dynamical symmetry breaking at the Planck scale. Unless one can eliminate or control the R term, the main point of our model is lost and one goes back to the conventional theory of gravity.

A second problem is that we have formulated the model in Euclidean space and for it to be interpreted we must transform back to Minkowski space. In most field theory models, this is a formality but in our model, because the signature of space-time is related to the compactness of the internal group, there might be a problem. Noncompact gauge groups can lead to nonunitary field theories. Whether this objection applies to our specific model remains to be investigated.

The model we present is closely related to the ideas of induced gravity.[6] The main difference between our approach and induced gravity is that induced gravity is due to all fields producing an effective Hilbert action, while in this model we single out a specific set of O(5) gauge fields as dominant and which have an algebraic structure that can be directly identified with the metric and connection. While still speculative, we believe these ideas are worth pursuing, especially as they suggest that gravitational effects could come in well below the Planck mass.

References

1. R. Utiyama, Phys. Rev. 101, 1597 (1956).
2. T.W.B. Kibble, J. Math. Phys. 2, 212 (1961).

3. S.W. MacDowell and F. Mansouri, Phys. Rev. Lett. 38, 739, (1977).
4. H. Pagels, Phys. Rev. D, (to be published 1983).
5. I thank William Marciano for this result.
6. S. Adler, Rev. of Mod. Phys., 54, 729 (1982).

PROGRAM

20th ANNUAL ORBIS SCIENTIAE

DEDICATED TO P. A. M. DIRAC'S 80th YEAR

MONDAY, January 17, 1983

Opening Address and Welcome

SESSION I-A:	GENERAL RELATIVITY AND GAUGE SYMMETRY
Moderator:	Behram N. Kursunoglu, University of Miami
Dissertators:	P. A. M. Dirac, Florida State University
	Behram N. Kursunoglu, University of Miami
SESSION I-B:	RELATIVISTIC COSMOLOGY
Moderator:	Michael Turner, University of Chicago
Dissertators:	Edward W. Kolb, Los Alamos National Laboratory "COSMOLOGICAL AND ASTROPHYSICAL IMPLICATIONS OF MAGNETIC MONOPOLES"
	Michael Turner, University of Chicago "THE INFLATIONARY UNIVERSE: EXPLAINING SOME PECULIAR NUMBERS IN COSMOLOGY"
Annotators:	Pierre Ramond, University of Florida
	Anthony Zee, University of Washington
SESSION II:	ROUND TABLE DISCUSSION ON DIRAC'S CONTRIBUTIONS TO PHYSICS
Moderator:	Abraham Pais, The Rockefeller University
Dissertators:	Harish-Chandra, Institute for Advanced Study, Princeton
	Fritz Rohrlich, Syracuse University "THE ART OF DOING PHYSICS IN DIRAC'S WAY"
	Victor F. Weisskopf, Massachusetts Institute of Technology

TUESDAY, January 18, 1983

SESSION III-A:	CURRENT STATUS OF EXPERIMENTS (W's, Z^o, MONOPOLE, PROTON DECAY)
Moderator:	Maurice Goldhaber, Brookhaven National Laboratory
Dissertators:	Frederick Reines, University of California, Irvine, and Daniel Sinclair, University of Michigan "A SEARCH FOR PROTON DECAY INTO $e^+\pi^o$ IRVINE-MICHIGAN-BROOKHAVEN COLLABORATION"
Moderator:	Joseph E. Lannutti, Florida State University
Dissertators:	Alfred S. Goldhaber, State University of New York at Stonybrook "MAGNETIC MONOPOLES AND NUCLEON DECAY"
Annotator:	P. A. M. Dirac, Florida State University
SESSION III-B:	ROUND TABLE DISCUSSION OF FUTURE ACCELERATORS
Moderator:	Alan D. Krisch, University of Michigan
Dissertators:	V. Soergel, DESY, West Germany "HIGH-ENERGY ep FACILITIES"
	R. R. Wilson, Columbia University "VERY-HIGH-ENERGY pp FACILITIES"
	Nicholas P. Samios, Brookhaven National Laboratory "HIGH-LUMINOSITY pp FACILITIES"
Annotator:	Ernest D. Courant, Brookhaven National Laboratory
SESSION IV:	COMPARISON OF THEORY AND EXPERIMENT
Moderators:	Sydney Meshkov, University of California, Los Angeles Pierre Ramond, University of Florida
Dissertators:	Sydney Meshkov, University of California, Los Angeles "GLUEBALLS"
	L. Clavelli, Argonne National Laboratory "ON THE MEASUREMENT OF α_s"
	Pierre Ramond, University of Florida "B-L VIOLATING SUPERSYMMETRIC COUPLINGS"
	Gabor Domokos, Johns Hopkins University "SPONTANEOUS SUPERSYMMETRY BREAKING AND METASTABLE VACUA"

Pran Nath, Northeastern University
"SUPERGRAVITY GRAND UNIFICATION"

Annotator: Richard Dalitz, University of Oxford

WEDNESDAY, January 19, 1983

SESSION V: FUNDAMENTAL PHYSICAL MECHANISMS OF BIOLOGICAL INFORMATION PROCESSING

Moderators: Gordon Shaw, University of California, Irvine
F. Eugene Yates, University of California, Los Angeles

Dissertators: Gordon Shaw, University of California, Irvine
"INFORMATION PROCESSING IN THE CORTEX: THE ROLE OF SMALL ASSEMBLIES OF NEURONS"

Robert Rosen, Dalhousie University, Canada
"INFORMATION AND CAUSE"

Michael Kohn, University of Pennsylvania
"INFORMATION FLOW AND COMPLEXITY IN LARGE-SCALE METABOLIC SYSTEMS"

George Wald, Harvard University
"LIFE AND MIND IN THE UNIVERSE"

Sidney W. Fox, University of Miami
"PHYSICAL PRINCIPLES AND PROTEINOID EXPERIMENTS IN THE EMERGENCE OF LIFE"

Annotator: Erich Harth, Syracuse University

THURSDAY, January 20, 1983

SESSION VI: FRAMEWORK OF ANALYSES OF BIOLOGICAL INFORMATION PROCESSING

Moderator: Michael Conrad, Wayne State University

Dissertators: Michael Conrad, Wayne State University
"MICROSCOPIC-MACROSCOPIC INTERFACE IN BIOLOGICAL INFORMATION PROCESSING"

Harold Hastings, Hofstra University
"STOCHASTIC INFORMATION PROCESSING IN BIOLOGICAL SYSTEMS II - STATISTICS, DYNAMICS, AND PHASE TRANSITIONS"

Otto E. Rössler, University of Tübingen, West Germany
"DESIGN FOR A ONE-DIMENSIONAL BRAIN"

SESSION VII:	THE HISTORY AND FUTURE OF GAUGE THEORIES
Moderator:	M. A. B. Beg, The Rockefeller University
Dissertators:	M. A. B. Beg, The Rockefeller University "DYNAMICAL SYMMETRY BREAKING: A STATUS REPORT"
	L. A. Dolan, The Rockefeller University "KALUZA-KLEIN THEORIES AS A TOOL TO FIND NEW GAUGE SYMMETRIES"
	G. Lazarides, The Rockefeller University "FLUX OF GRAND UNIFIED MONOPOLES"
	S. Mandelstam, University of California, Berkeley "ULTRA-VIOLET FINITENESS OF THE N = 4 MODEL"

FRIDAY, January 21, 1983

SESSION VIII:	GENERAL RELATIVITY IN ASTROPHYSICS
Moderators:	P. G. Bergmann, New York University Joseph Weber, University of Maryland and University of California, Irvine
Dissertators:	Giorgio A. Papini, University of Regina, Canada "GRAVITATION AND ELECTROMAGNETISM COVARIANT THEORIES A LA DIRAC"
	Joseph Weber, University of Maryland and University of California, Irvine "GRAVITATIONAL WAVE EXPERIMENTS"
	Kip S. Thorne, California Institute of Technology "BLACK HOLES"
Annotator:	John Stachel, Boston University
SESSION IX:	GENERAL RELATIVITY IN PARTICLE PHYSICS, QUANTUM GRAVITY
Moderator:	Anthony Zee, University of Washington
Dissertators:	Anthony Zee, University of Washington "REMARKS ON THE COSMOLOGICAL CONSTANT PROBLEM"
	John Schwarz, California Institute of Technology "A NEW FORMULATION OF N=8 SUPERGRAVITY AND ITS EXTENSION TO TYPE II SUPERSTRINGS"
	Heinz Pagels, The Rockefeller University "GRAVITATIONAL GAUGE FIELDS"

PARTICIPANTS

Ahmed Ali
DESY, Hamburg, Germany

M. A. B. Beg
The Rockefeller University

Carl M. Bender
Washington University

Peter G. Bergmann
Syracuse University

John B. Bronzan
Rutgers University

Arthur A. Broyles
University of Florida

Roberto Casalbuoni
Istituto di Fisica Nucleare
Florence, Italy

Lay Nam Chang
Virginia Polytechnic Institute
and State University

L. Clavelli
Argonne National Laboratory

Michael Conrad
Wayne State University

John M. Cornwall
University of California
Los Angeles

Ernest D. Courant
Brookhaven National Laboratory

Thomas Curtright
University of Florida

Richard H. Dalitz
Oxford University
Oxford, England

Ashok Das
University of Rochester

Stanley R. Deans
University of South Florida

P. A. M. Dirac
Florida State University

Louise Dolan
The Rockefeller University

Gabor Domokos
Johns Hopkins University

Susan Kovesi-Domokos
Johns Hopkins University

Bernice Durand
University of Wisconsin

Loyal Durand
Fermi National Accelerator
Laboratory

Robert W. Flynn
University of South Florida

Sidney W. Fox
University of Miami

Andre I. Gauvenet
Electricite de France, Paris, France

Alfred S. Goldhaber
State University of New York

Gertrude S. Goldhaber
Brookhaven National Laboratory

Maurice Goldhaber
Brookhaven National Laboratory

O. W. Greenberg
University of Maryland

Franz Gross
College of William and Mary

Bernard S. Grossman
The Rockefeller University

Gerald Guralnik
Brown University

M. Y. Han
Duke University

Harish-Chandra
Institute for Advanced Study

Erich Harth
Syracuse University

Harold Hastings
Hofstra University

K. Ishikawa
City College of the CUNY

Gabriel Karl
University of Guelph

Boris Kayser
National Science Foundation

Michael S. Kohn
University of Pennsylvania

Edward W. Kolb
Los Alamos National Laboratory

Alan D. Krisch
University of Michigan

Behram N. Kursunoglu
University of Miami

Joseph E. Lannutti
Florida State University

G. Lazarides
The Rockefeller University

Y. Y. Lee
Brookhaven National Laboratory

Don B. Lichtenberg
Indiana University

Stanley Mandelstam
University of California
Berkeley

Philip Mannheim
University of Connecticut

Jay N. Marx
Lawrence Berkeley Laboratory

Koichiro Matsuno
University of Miami

Sydney Meshkov
University of California
Los Angeles

A. J. Meyer II
The Chase Manhattan Bank, N. A.

Stephan L. Mintz
Florida International University

John W. Moffat
University of Toronto

Paul Mueller
University of Pennsylvania

Darragh Nagle
Los Alamos National Laboratory

Pran Nath
Northeastern University

John R. O'Fallon
University of Notre Dame

Heinz R. Pagels
The Rockefeller University

Abraham Pais
The Rockefeller University

William F. Palmer
Ohio State University

Giorgio A. Papini
University of Regina
Regina, Canada

Arnold Perlmutter
University of Miami

Pierre Ramond
University of Florida

L. Ratner
Brookhaven National Laboratory

Paul J. Reardon
Brookhaven National Laboratory

Frederick Reines
University of California
Irvine

Fritz Rohrlich
Syracuse University

Robert Rosen
Dalhousie University
Halifax, Canada

S. Peter Rosen
National Science Foundation

Otto E. Rössler
University of Tübingen
Tübingen, Germany

Nicholas P. Samios
Brookhaven National Laboratory

Mark Samuel
Oklahoma State University

John H. Schwarz
California Institute of
Technology

Joel Shapiro
Rutgers University

Gordon Shaw
University of California
Irvine

P. Sikivie
University of Florida

Daniel A. Sinclair
University of Michigan

George A. Snow
University of Maryland

V. Soergel
DESY, Hamburg, Germany

John Stachel
Boston University

Jerry Stephenson
Los Alamos National Laboratory

Katsumi Tanaka
Ohio State University

Michael J. Tannenbaum
Brookhaven National Laboratory

L. H. Thomas
National Academy of Sciences

Kip S. Thorne
California Institute of Technology

Michael S. Turner
University of Chicago

Giovanni Venturi
Universita di Bologna
Bologna, Italy

George Wald
Harvard University

Kameshwar C. Wali
Syracuse University

Joseph Weber
University of Maryland

Victor F. Weisskopf
Massachusetts Institute of
Technology

Robert R. Wilson
Columbia University

F. Eugene Yates
University of California
Los Angeles

G. B. Yodh
University of Maryland

Frederick Zachariasen
California Institute of
Technology

Cosmas K. Zachos
Fermi National Accelerator
Laboratory

Anthony Zee
University of Washington

INDEX